THE MEMORY OF WATER

MICHEL SCHIFF

THE MEMORY OF WATER

HOMOEOPATHY AND THE BATTLE OF IDEAS IN THE NEW SCIENCE

Thorsons
An Imprint of HarperCollins*Publishers*

Thorsons
An Imprint of HarperCollins*Publishers*
77–85 Fulham Palace Road,
Hammersmith, London W6 8JB
1160 Battery Street,
San Francisco, California 94111–1213

Published by Thorsons 1995
1 3 5 7 9 10 8 6 4 2

A catalogue record for this book
is available from the British Library

ISBN 0 7225 3262 8

Typeset by Harper Phototypesetters Limited,
Northampton
Printed in Great Britain by
HarperCollinsManufacturing Glasgow

CONTENTS

FOREWORD

WHEN the editor asked me to write the foreword to the English version of Michel Schiff's book (*Un cas de censure dans la science: L'affaire de la mémoire de l'eau*) my first move was to turn down this proposal: there was enough of 'Benveniste' in this book. But then the author succeeded in convincing me that the reader could be interested to know what I had to say about how an external observer viewed my work. So I agreed. When the time came to write this introduction, I initially decided to comment not on the book but on what was lacking in it. Indeed, when I first read the book I thought I would not be interested by this account of a series of events that I should know better than anyone else. To my surprise this was not so. I then realized that not only had I forgotten many of the episodes of the saga but also that Michel Schiff has seen, from both his privileged external point of view and his expertise in the sociology of science, many aspects of the struggle that did not appear to the daily participants, including myself. Furthermore he brought to light what we vaguely knew but never clearly formalized, that is, the enormous number of positive experiments, done blind or in clear, that we had performed over time on the two main systems that are the basis for the controversy: the high dilution and the transmission experiments.

I found in this book what I expected: a clear diagnosis of the reluctance of the scientific community to change paradigms and an account of how the so highly celebrated peer-review system (paraphrasing Churchill: 'the *pire* [French for 'worse'], indeed, at the exception of all others') went berserk when confronted with results that referees could not read, using the 'software' they have in hand, that is, the currently prevailing paradigm. What is also clear is that many authors starting from Kuhn have proposed theories to explain such reluctances to shift paradigms. Yet, when the crisis occurred, it was as if it were the first one in the history of science and all the *lieux communs* and errors of judgement were mouthed and printed, the following being exemplary: 'It cannot be since if it were true it would have been found two hundred years ago' . . . 'It cannot be because it would nullify years, centuries of knowledge' . . . 'It cannot be because it is not always reproducible' . . . 'It cannot be because there is no theory behind it', and so on. On this deaf and blind attitude, this book is in fact a textbook.

However, as an experimenter, that is, a person who likes to find solutions to yet unsolved problems, I would like to express two concerns that are not to be found here. The first is that, when I read in *Nature* remarks about unusual results requiring 'different editorial standards' (Henry Metzger), it seems to me that such an opinion (which unfortunately is accepted by most scientists) explains by itself the impressive slowing down of science. Biology still lives according to a nineteenth-century paradigm, and the modern revolution of physics has seemingly been achieved in the thirties. (For those who *volens nolens* appear to pay tribute to the God Progress, it should suffice to say that there has been *no* paradigm shift in the time between Mendel's experiment and the cloning of a gene via molecular biology techniques; the double helix, as important as this discovery is, is simply the structural materialization of what was known as chromosomes and genes, and the cloning process is purely technological.) Obsessed with 'quality' and big science, the scientific system has built an instrument which indeed insures 'quality' but strictly within the admitted paradigm since, by definition, the judges belong to the existing system; when confronted with any data that could represent such a change, they react by demanding special laws to deal with special results. In fact, should the system be intellectually regulated, a completely opposite stance would be taken; that is, fearing the deleterious effect of the self-perpetuating reviewing system, we should heartily build a counter system that would be lenient to controversial, weak, premature data. One can immediately see the main objection to this – how can this system be made foolproof to errors or even fraud? This objection has in fact no interest whatsoever, for the simple reason that scientists do not create nature, they only 'dis-cover' it, in other words, removing the cover under which nature is provisionally hidden. There is no way, and especially since science is now such a large industry with plenty of means at hand, that false or fraudulent data can stand very long before being disproved. (In passing, one of the arguments in favour of the existence of what is known as the memory of water is the fact that, after 10 years, no one has come up with a decent alternative explanation for these data.) If science does not nurture its infant results with the same care as human beings nurture their fragile offspring, if this kind of 'special law' akin to the most autocratic regimes prevails, then one can easily predict that science will not survive this final blow.

My second concern, which is especially valid for biology for the simple reason that biology deals with the most complex set of systems that is life, is that it is now apparent to most scientists (even if they don't phrase it loudly) that the current process of reductionism, of microanatomy, that is summarized in the statement: 'When we know all the parts we will know how the whole works' is wrong. This is because the amount of information that becomes quickly available a few years after the opening of a new field (e.g. research into interleukins or the protein kinases) is so large that no human mind can understand not only the role of each item but much less their multiple interactions. This is why, after decades of effort, we understand practically no better the response and the control of the immune system and how and why a cell becomes cancerous, and have, as a consequence, practically no better means than we had 30 years ago to control these processes. The time has come now where this gap between the promises of scientists and what they have actually fulfilled to help their fellow tax-payers will become apparent even to the most faithful or blindest politician. By coincidence, it might well be that what we have recently unveiled, that 'molecules communicate via specific electromagnetic waves', might give us a tool to tackle biological systems no longer from the point of view of the structure, which is an endless, one-by-one dissection process, but by using modern electronic and computerized means. This brings us back to the memory of water and to the opening of new possibilities to fulfil these promises and trigger one of these rare paradigmatic shifts in biology (and possibly also in the physics of condensed matter). In this, there is still a long fight in front of us, but I believe this book may represent a decisive step towards this victory.

J. Benveniste 1995

INTRODUCTION

'HOMOEOPATHIC dilutions' and 'memory of water' are two expressions capable of turning a peaceful and intelligent person into a violently irrational one. This book is a testimony of scientific studies of homoeopathic dilutions and other related phenomena, but it also deals with the way in which these studies have so far been received by most scientists. These experiments and the reactions to it have led to a long-running scientific dispute known as the 'Benveniste affair'.

One of the teachers in my former secondary school used to tell his students that he did not believe in atoms. I myself was fortunate enough to have a more enlightened teacher, who communicated to me his enthusiasm for science; on the subject of atoms, I remember that he used to say: 'How can anyone deny their existence? Nowadays, we can count them, we can weigh them, we can even take them for a walk!'

Perhaps this phrase played a role in my decision, 20 years later, to translate into French a classic textbook of atomic physics that I had used as a graduate student.[1] It may even have contributed 40 years later to my curiosity about reports that seemed to contradict the idea that atoms and molecules are the basis of chemical and biological interactions. As is well known, homoeopathic medicine frequently uses solutions so highly diluted that no molecules of the original active substance should be left to act chemically or biologically.

In the 1980s a well-established scientist provided evidence for the most controversial aspect of homoeopathy by confirming the ability of water to 'remember' previous contacts with biologically active chemical; Jacques Benveniste is the French scientist who directed the team that achieved several of the breakthroughs in this study. After 4 years of work on high dilution experiments, his team then published an article in *Nature*, the most influential of all scientific journals. The reactions to that publication were very violent. In particular, within a few days,

Nature's editor-in-chief came to Benveniste's laboratory together with a physicist specializing in the detection of scientific fraud and a professional magician. Shortly after this 5-day visit, this scientific 'commando unit' published a report in *Nature* entitled ' "High dilution" experiments a delusion'.

In a book concerning plane accidents, a French pilot once wrote: 'If a pilot tells you that he never makes mistakes, take my advice and don't fly with him, because he is dangerous.' At the beginning of this journey to the frontiers of science, I can say the same about scientists. If a scientist tells you that his sole guide is 'objectivity', that he is only seeking 'scientific truth' as it is revealed by 'facts', then be careful: he is lying to himself, and he will mislead you in good faith. I am a scientist, yet I do not claim that my testimony about the memory of water and about scientific censorship is completely 'objective'. Like any other person, I have feelings which sometimes interfere with my judgements. Nevertheless, I hope that an awareness of these feelings has helped me to develop a better understanding of the scientific facts concerning the memory of water and to acquire some insight into the human aspects of these facts. For instance, I became quickly aware of the danger of identifying too closely with Benveniste, which might lead either to fantasies about a Nobel Prize by proxy or to a confusion between his conflicts with the establishment and mine. Hence I spent only 2 days a week in his laboratory, keeping the rest of the time for reading and for interactions with other scientists.

In this book, I describe the results of 3 years of investigation into the 'memory of water', a phrase which will be used in this book to designate the strange behaviour of water molecules, which somehow seem able to keep a record of previous contact with other kinds of molecules. The behaviour of many fellow scientists has been equally strange. As I will try to show, the moral of the story is that, in a scientific controversy, human factors are thoroughly intertwined with the more evident technical factors.

Since I am stressing the human aspects of scientific knowledge, I might as well say a few words about myself as a scientist. I was trained as a physicist and received my PhD from the University of Chicago, where I worked in high energy physics. When I was a graduate student, research in high energy physics could still be carried out on a small scale. However, this field of research has now turned into a big business, using increasingly large instruments, and I became disenchanted with it. The next area of research which I explored – the IQ controversy and other questions concerning genetics and human behaviour – seemed to me more stimulating and socially more important; I spent 15 years working with it.

My work in this field made me aware of the importance of human prejudice in scientific research; I have since analysed the issue of subjectivity in science in other areas of research. Over the last two decades, social studies of science have shown that human factors are important even in the so-called 'hard' sciences such as physics and chemistry. The lay person may not worry too much about the consequences of academic prejudice on nineteenth-century chemistry or on contemporary astrophysics. In the case of the memory of water, however, the issues are less academic; they bear directly on medical research, on human health and on the experts' monopoly on medical knowledge. This is why I have tried to write my account in a non-academic manner. But, because the issues are also technical in nature, technical aspects must also be examined, and the details of these can be found in the Appendixes.

I became interested in the memory of water in 1988. I even wrote 10 pages about it in a book that was published in French in March 1992. In it, I took a sceptical position on the reality of the phenomenon. At the same time, I was critical about what appeared to me to be a case of scientific censorship. My direct involvement with Benveniste's research on the memory of water started in March 1992. In exchange for an opportunity to study his research on the memory of water, I occasionally took part in his latest experiments on that topic. I became so intrigued that I finally participated more directly in some further 'transmission' experiments, in which chemical information seems to be transmitted through an electronic device without the concomitant transport of matter.

During the 3 years of my investigation, I have tried to understand both the scientific and the social aspects of the conflict. While doing so, I witnessed many examples of scientific irrationality. I do not claim to have learned all there is to know about the matter, but information about research on the memory of water and about scientific censorship is so sorely lacking that I felt a sense of urgency. I hope that my testimony will be useful to those who want to know more about the scientific status of homeopathic dilutions and about what happens when scientific orthodoxy appears to be threatened.

Benveniste makes claims which are challenging both from the scientific and from the sociological point of view. Scientifically, he describes observations which cannot be explained by current theories. He also points out that it is precisely this lack of an adequate theory which makes his experiments interesting, because they should spur scientists to re-evaluate their current knowledge. From a sociological point of view, the adamant refusal of scientists to enter into a serious discussion is an indication that there is something rotten in the kingdom of Academia. Whilst I do not analyse the resistance of scientists to new ideas in exactly the same way as Benveniste does, I do agree that there are serious signs of censorship and of self-censorship.

The 'trees' of the Benveniste affair have tended to hide the 'forest' of the memory of water. It is much easier to attack Benveniste as a person and as a scientist (or to defend him) than to reach a reasoned opinion about the memory of water and about censorship in science. According to a Chinese saying, when a man is pointing to the moon, the fool looks at the finger while the wise person looks at the moon. We shall try to be wise.

PART ONE

THE STRANGE BEHAVIOUR OF ORDINARY WATER

CHAPTER ONE

ANOMALIES OF ALL DISCIPLINES, UNITE!

SEVERAL groups of scientists have now reported on the biological effects of 'high dilutions'. In this chapter, I provide the reader with a theoretical perspective on these effects. The purpose is to show that the biological activity of some homoeopathic dilutions observed by various scientists is in no way a pathological or isolated fact. On the contrary, it is only one of a series of observations that challenge current views about the interaction between water, electromagnetic fields and living cells.

THE PUZZLE OF HOMOEOPATHIC DILUTIONS

The idea that matter has a discontinuous, granular structure is very old. Etymologically, the word 'atom' means 'that which cannot be divided'. Some Ancient Greeks asked what would happen if one tried to divide matter into increasingly smaller pieces; would one finally reach a situation where each piece of matter would be like a point and could no longer be divided? Until the development of nineteenth-century chemistry, however, these questions were nothing but philosophic speculations.

As we all learned in secondary school, atomic theory now accounts for a large number of empirical observations. For instance, it explains the law governing the weights of chemical elements that combine into a given compound. If one were to repeat many times Lavoisier's experiment combining hydrogen with oxygen to produce water, the proportion of oxygen and hydrogen would always be exactly the same: 8 grams of oxygen to each gram of hydrogen. These fixed proportions are accounted for by the fact that, while 2 atoms of hydrogen combine with 1 atom of oxygen, as indicated by the chemical formula H_2O, the 1 atom of oxygen has 16 times the weight of 1 atom of hydrogen; therefore the ratio of weights is 8 to 1.

At school we also learned about Avogadro's constant; this constant (6×10^{23}) denotes the number of molecules that always occurs in 1 mole

(SI unit of amount) of substance. Therefore 1 gram (1 mole) of hydrogen and 16 grams (1 mole) of oxygen contain the same number of atoms. This number is gigantic (roughly one million billion billions), but it is not infinite. Let us consider for instance 18 grams (1 mole) of water, containing $N = 6 \times 10^{23}$ water molecules. Now imagine dividing them by 10, and again by 10, and again by 10, and so on. If you did this 23 times, you would finally end up with stacks each consisting of 6 molecules, and at that point it would be impossible to divide by 10 any further.

The repeated division by 10 corresponds to the way decimal homoeopathic dilutions are prepared. Suppose that a small quantity of some active chemical substance containing 10^{12} molecules was dissolved within a given volume of water. A decimal dilution is obtained by mixing one-tenth of the previous solution with nine times that volume of pure water, to arrive back at the original volume. Each volume of the first decimal dilution would then contain one-tenth of the original molecules of the active substance (namely 10^{11} molecules). If one were to repeat this dilution 11 more times, each volume of the last of these dilutions would, on the average, contain only one single molecule of the active chemical product. Repeating the dilution process yet again, the liquid would then contain no molecules of the active substance. For instance, in the 20th decimal dilution one would have to examine 100 million samples to find one single molecule of the original chemical! The enigma of the biological effects of such dilutions relates to the medical action of high-potency homoeopathic remedies – that is, how does a small volume of liquid seem to act on a biological cell in spite of the fact that it contains no molecules of the active chemical?

Results of scientific tests of extreme dilutions seem in complete contradiction to modern ideas about the atomic, discontinuous structure of matter. When faced with this apparent contradiction, most scientists, instead of saying 'We don't understand, let us seek an explanation', react by saying 'We don't understand, therefore it is impossible.' By doing so, they ignore all previous examples in the history of science where an apparent anomaly finally provided important new insights into natural phenomena.

In the second part of the book, we shall examine the various strategies used by scientists to avoid asking: 'What if it were true?' One of these strategies has been to raise the spectre of scientific chaos, as if experiments on high dilutions could be taken seriously only by discarding two centuries of scientific research. The truth of the matter is that no scientist studying high dilutions ever talked of discarding atomic theory, or challenged Avogadro's number. What scientists studying high dilutions have said is: 'Here are some observations that we don't understand; let us look for an explanation.'

Actually, the hypothesis known as the memory of water does not imply negating the existence of atoms and molecules, but rather the capacity of water molecules somehow to organize in a stable manner and through such an organization to acquire the capacity of storing information obtained from other molecules. This stored information could then be played back, like a symphony that has been recorded on a magnetic tape. In this perspective, the molecular organization of water is not negated but enriched.

To the best of my knowledge, as of today, no scientist claims to have a working theory of the memory of water. Nobody knows precisely how the molecular organization of water could permit the storage and playback of chemical information. It is this lack of understanding which is a challenge to scientists. In the remainder of this chapter, I will show that the memory of water is only one of the various signs that there is something fundamentally inadequate in our current understanding of water as a liquid, as a chemical solvent and as a component of biological cells.

BIOMAGNETISM: FROM POPULAR TO SCIENTIFIC KNOWLEDGE

Like the word 'energy', the word 'magnetism' is used in different ways by scientists and by lay people. For scientists, magnetism refers to the properties of magnets, to magnetic fields and to the effects of these fields on matter. A more popular use of the word 'magnetism' refers to human beings; the association of electromagnetism and living beings has been a highly controversial subject, smacking of 'vitalism', and until recently forbidden territory to scientists. The scientist who dares to trespass over the line drawn by the establishment around such a territory is automatically excommunicated.

An example of such an excommunication is that of the French physicist Yves Rocard. After a successful career that led him from the construction of the French nuclear bomb to the head of the prestigious Ecole Normale Supérieure, he became interested in the sensitivity of the human body to magnetic fields. Through his research, he was trying to provide a scientific explanation for what he had been able to observe about water diviners.

In this particular case, it is not his answer that was unacceptable to many scientists, since it was quite conventional: he attributed this sensitivity to magnetic fields to small crystals of magnetic material within the body. What was forbidden was the question itself. On the one hand, studying how animals orient themselves or communicate through some physical means like ultrasonic sound is considered acceptable scientific research; on the other, applying the same methods to the study of the human body has often resulted in accusations of mysticism or fraud because of the existing dogma that electromagnetism could not possibly have any role in humans.

This situation has recently changed, however. I do not mean that you could obtain a grant from the US National Science Foundation to study people looking for hidden springs with wooden sticks. But, provided that you used the proper scientific decorum, you could now study the interaction of living matter with electric and magnetic fields without being expelled from the scientific community. What probably forced scientists to abandon their ostrich-like behaviour concerning the interaction between magnetic or electric fields and living matter was popular pressure over health hazards associated with various types of electrical devices, especially power lines. Whatever the reason, the status of research on biomagnetism is no longer that it is an area reserved for lunatics and charlatans, but a respectable (though still controversial) topic of scientific investigation.

In several countries, scientists have started to study the effects of various types of industrial fields on the development of certain forms of cancer. Others have examined the effects of such fields on biological cells in test tubes in the laboratory. Two facts indicate that biomagnetism is becoming a respectable topic of scientific research: the first is that learned societies are now holding professional meetings about this type of research; the second is the appearance of review articles in journals – a phenomenon which reveals a growing proliferation of research reports on a given topic. In the year of 1992 alone, the FASEB journal (one of the leading biology journals) published four review articles on the various effects of periodic fields on the immune system. The effects of low frequency electric or magnetic fields are now being studied in many laboratories and a recent international conference on that topic attracted a thousand participants. The fact that low frequency fields can influence the development of living cells is the first of a series of scientific puzzles that may be related to that of the memory of water.

Like the biological activity of high dilutions, this scientific anomaly is situated at the border between physics, chemistry and biology. Here once again, scientists are faced with an apparent 'impossibility'. However, in the case of bioelectricity and biomagnetism, the fact that there really is something to be investigated is accepted by an increasing number of biophysicists. The reality of the phenomena is now accepted, yet this reality still creates a scientific enigma, because fields of very low frequencies should in principle have no measurable effects.

In order to give the reader a feeling for the theoretical problem, I will use an analogy with familiar phenomena. If you take a small piece of iron like a tuning fork, it can oscillate several hundred times per second. If you build a suspended bridge of the same material, it can oscillate with a frequency of about once per second. When I was in the army, I learned that we were not allowed to march over such a bridge because of the risk of resonance; if the frequency of our steps happened to match that of the bridge, the vibrations could be strong enough for the bridge to collapse. I am trying to illustrate two points with this analogy. The first is that the frequency of vibration of an object decreases as its size increases. The second is that you can modify the structure of an object if you subject it to the appropriate frequency of vibration.

The frequencies of atoms or molecules are very much higher than the 50 or 60 periods per second of industrial fields; in principle, such low frequency fields should have no influence on individual atoms or molecules. The analogy with the tuning fork and the bridge suggests, however, that perhaps low frequency fields act not on single molecules but on larger objects containing a great number of molecules. The anomalous effects of low frequency fields on living cells mean that the usual rules governing the interaction between fields and matter (described by quantum mechanics) may need to be changed, at least when atoms or molecules act in a collective manner, as they probably do within the water of biological cells. A theory recently put forward by two Italian scientists, Emilio Del Giudice and Giuliano Preparata, and known as the 'theory of coherent domains', predicts such collective behaviour of water molecules (*see later in this chapter*).

HOW DO MOLECULES COMMUNICATE WITHIN A LIVING CELL?

Scientists rarely ask embarrassing questions – that is, those questions for which they don't already know most of the answers. The question posed in the title of this section is one which scientists tend to shun. Nevertheless, a few have been curious about the way chemical processes occur within a living cell. Benveniste is one. In a paper written for a residential seminar which I organized with him, he asked the following question: 'How can we understand that molecules meet within a cell if we remember that the universe of the cell is gigantic on a molecular scale?' This question can be reformulated as two distinct sub-questions: (1) how do specific molecules manage to communicate at a distance and find out that they are capable of interacting? and (2) what force steers them together for this interaction?

This brings us to the idea of action at a distance between molecules. As we shall see, this is the key concept of the theory of coherent domains of the Italian physicists. Because of the vast difference between the sizes of molecules and of biological cells, this question of action at a distance between molecules cannot be avoided. If you take this question seriously, you have to assume that molecular messages can travel through intracellular space, which is largely filled with water – in other words that, at least within a biological cell, *the chemical information of a molecule can travel through water.*

The idea of action at a distance is the basic idea of field theory, but it contradicts our everyday intuition, because familiar objects interact only when in close contact. In this respect, we are almost as poorly equipped to accept the idea of action at a distance as were the predecessors of Newton, who had to think of angels pushing and guiding the planets along their orbits in order to avoid considering that idea. If you reject action at a distance between molecules, however, you have to rely solely on thermal agitation to bring them in close contact in the appropriate way; this may be as inadequate as ignoring the impact of radio and television on the sociology of human interactions.

In conclusion, I want to emphasize how important water is in biology. To stress this importance, the Nobel Prize winning biochemist Szent-Györgyi used to say humorously that proteins (the building blocks of living matter) are nothing but impurities of water. Although this is of course an exaggeration, it serves to remind us of the fact that, within a living cell, each molecule of protein is surrounded by thousands of molecules of water. In the next century, water will perhaps be considered more significant than DNA for an understanding of life.

Like questions about the memory of water, the question of molecular communication within cellular water concerns molecular information leaving its original basis within the molecule. However, the problem posed by high dilution experiments is even more puzzling, because the information must be able not only to travel through water but also to be stored by it. The theory of coherent domains outlined in the next section might provide a model for this.

COHERENT DOMAINS: NEW PHYSICAL AND CHEMICAL OBJECTS?

Giuliano Preparata and Emilio Del Giudice are two scientists working at the Milan Institute for Nuclear Physics. Formerly associated with CERN (the European Centre for Nuclear Research), Preparata now holds the chair of nuclear physics at the University of Milan. These two Italian physicists are engaged in an ambitious programme of research in an attempt to explain some anomalies of liquids and solids. Here, I will limit my presentation to the physics of water, and particularly the way temperature influences its properties.

Water is not only an essential ingredient of life; from a purely physical point of view, it is also a substance with many anomalies. One anomalous property is its ability to remain liquid at relatively high temperatures. Another is the low density of the solid state of water (ice), to the extent that it floats on top of the liquid state. Both these anomalies contributed to the appearance of life on Earth. Others of biological importance include the abnormally high value of its dielectric constant; it is this which permits very large electric fields within living cells.

In the nineteenth century, physicists were very successful in understanding the overall behaviour of gases, which act essentially as assemblies of *isolated* points. Each of these 'points' is in fact an atom or a molecule with a complex structure. In the twentieth century, physicists have developed their understanding of these complex structures using the theory of quantum mechanics, which can account for many of the properties of individual atoms and individual molecules. Its ability to explain the periodicity of the properties of chemical elements as summarized by Mendeleev's table has been most impressive. However, it is less successful in explaining what happens when atoms or molecules get so close that the gas becomes a liquid or solid.

Contemporary models of condensed matter are a patchwork composed of the physics that was successful in explaining the behaviour of gases (the so-called classical mechanics) and the physics that was successful in understanding the structure of individual building blocks (the so-called quantum mechanics). These models break down, however, when the density of atoms or molecules is a hundred to a thousand times greater than that found in gases. In a gas, individual 'points' are far from each other; apart from individual collisions where they bounce like miniature billiard balls, these 'points' have no significant interaction. When atoms or molecules are closely packed, however, the separation between what is going on outside the structure (classical mechanics) and what is going on inside (quantum mechanics) is no longer valid. In the absence of an overall theory to account for this situation, physicists have used various models based on a large number of empirical parameters.

A comparison with the historical development of astronomy illustrates the distinction between working models and genuine theories. For more than a thousand years, astronomers were able to predict the motions of visible planets, including the timing of eclipses, with reasonable accuracy by using a very complicated model proposed by the Egyptian astronomer Ptolemy. This model accounted for most of the empirical observations able to be made at the time, in spite of the fact that its basic assumption was wrong – that is, that the Sun and the planets revolved around the Earth. Galileo's observations of the phases of Venus was a powerful argument in favour of a heliocentric (sun-centred) model. Even so, the former model could probably have been saved at the time by adding more assumptions and parameters to it. The transition from a model to a genuine theory was finally provided by Kepler's laws governing the motion of planets and by Newton's theory of universal gravitation. After this radical change in astronomical theories, predictions did not become dramatically better, however, and few novel phenomena were predicted. The essential improvement was that a vast number of observations could now be *related*; for instance, the gravitational constant is the same for all planetary motions and these motions can be related to other phenomena like the motion of a pendulum, the fall of an apple, and the motion of water known as tides.

The state of knowledge about liquids and solids (and in particular about water) is like the state of knowledge of astronomy before Galileo, Kepler and Newton. This state of affairs can also be illustrated by the following quip from workers in the field of computer modelling: 'Give me three parameters and I will draw an elephant; give me four and I will make it walk', a phrase which serves to emphasize their basic ignorance. The goal of the research programme of Preparata and Del Giudice is to relate the vast amount of empirical data on liquids and solids to a few basic facts instead of remaining satisfied with the unrelated and *ad hoc* models currently existing.

In March 1993, Emilio and Giuliano came to Paris for the 3-day seminar which I had organized with Benveniste. One day, they happened to enter a bookstore where the French Nobel physicist De Gennes was signing his book. With his charming foreign accent, one of them asked: 'Professor De Gennes, how can the wings of a plane stay solid? How do the atoms stay properly aligned if they can "feel" only their nearest neighbours?' In a deceptively simple manner, the Italian physicists were asking one of those embarrassing questions which scientists tend to avoid. The paradox raised by this apparently naïve question is the following: if atoms were held together only by short range forces, solid matter would probably be unstable, like a castle built of cards. According to Preparata and Del Giudice, stability can be achieved only through the existence of long range forces, which lead to collective behaviour. These are also needed to account for the existence of liquid water: such forces make it energetically more attractive to molecules to be in the condensed form of a liquid rather than in the more dispersed form of a vapour.

Del Giudice and Preparata may appear as revolutionaries. In fact, they are faithful adepts of the most basic theory of modern physics, namely quantum mechanics. According to their theory, packed molecules form coherent domains because such an organization results in lower energy levels. Molecules spontaneously go from a chaotic state to an ordered state if that ordered state happens to contain less energy. The principle involved is that the stable state of a system is the one with minimal energy. (It is perhaps worth stressing that this principle is one of the most fundamental principles of physics; it is considered to be valid both in classical physics and in quantum mechanics.)

To return to our water molecules, in isolated atoms or isolated molecules the only relevant forces are the short range, electrostatic ones, which are analogous to hooks. These forces become negligible beyond the nearest neighbours. In quantum mechanics, however, once atoms or molecules are sufficiently packed the long range forces start to play a significant role because, although these forces are very weak, they operate over greater distances. Hence they modify the energy balance in a significant manner when there are more particles per unit volume. This modification of the energy balance leads to qualitatively new effects; Preparata and Del Giudice have shown by exact computations that atoms or molecules when closely packed have a collective behaviour and act coherently, as a whole (forming 'coherent domains'), rather than as isolated objects. Metaphorically speaking, they are 'marching together' instead of wandering about in a chaotic manner, in the case of liquid water forming domains containing millions of molecules. This theory enabled Preparata and Del Giudice to explain some observations of the behaviour of liquids and solids, including some bizarre and previously inexplicable ones (e.g. where the absolute temperature scale, starting at absolute zero, seems to shift upwards). To end this sketchy outline of the theory of coherent domains, I have reproduced a table of some of the quantitative results derived so far from this theory (Table 1.1).

TABLE 1.1 *Predictions and observations using the theory of coherent domains*

	Numerical values	
	Predicted	*Observed*
Water		
Critical volume of N molecules	57cm³	55.6 cm³
Second Van der Waals coefficient	4.4	5
Latent heat	9.8 Cal/mol.	9.7 Cal/mol.
Helium		
Temperature of phase transition	2.25 K	2.18 K

In his influential book *Les Atomes*, Perrin argued that atoms apparently did correspond to some underlying reality since their number in a given quantity of matter could be calculated in 13 different ways, all leading to the same value. (This value was obtained from measurements concerning phenomena as different as the colour of the sky, the viscosity of gases and the erratic (Brownian) motion of small particles floating in liquids.) The results in the table above are a long way still from the 13 coincident results noted by Perrin with respect to atoms, so that coherent domains cannot yet claim this status of 'underlying reality'. However, if successful correspondences between predictions and observations continue to accumulate, the situation may change before the end of the century. The question posed by the theory of coherent domains is whether scientists will enter the twenty-first century working from ideas of classical physics developed during the nineteenth century, or whether they will finally integrate the totality of quantum field theory developed in the twentieth.

'I DON'T WANT TO HEAR ABOUT IT'

I noted in the Introduction that the trees of the Benveniste affair have been hiding the forest of the memory of water. In this chapter, we have seen that the issue is actually wider than that of homoeopathic dilutions. The memory of water appears to be only one piece of a larger scientific puzzle. For the moment, it is impossible to know what the precise relationship is between the various phenomena described here: the biological potency of homoeopathic dilutions, chemical signals at a distance, the action of low frequency fields on some cellular processes and the various anomalous physical properties of water.

The connection between these phenomena may finally turn out to be quite different from that suggested by the theory of coherent domains. The history of science shows that hidden links between phenomena become evident only when the theoretical problems posed by these phenomena have been solved. The history of the long development of atomic theory and of quantum mechanics provides a good illustration of this point. Before the full development of atomic theory, nobody could have known that there was a close link between the colour of the sky, the viscosity of gases and the chaotic motion of small particles suspended in liquids. Similarly, before the full development of quantum theory, who could have guessed that the analysis of the radiation coming out of an oven (the so-called 'black-body' radiation) would lead to a theory providing an explanation of the chemical regularities in the elements which had been discovered half a century earlier by the Russian chemist Mendeleev?

A well-known historical example illustrates what can be learned from the past about hidden links between phenomena. This example concerns the age of the Earth. At the end of the nineteenth century, geologists evaluated its age as being of the order of magnitude of one billion years. To do this, they used observations and theories specific to their field. Geology was low in the pecking order of scientific disciplines so that this evaluation was not taken seriously when it was contested by physicists. At the time, physics was paramount among the scientific disciplines, as is still the case. Lord Kelvin, one of the leading physicists of his time, stated that the age of the Earth estimated by geologists was nonsense.[1]

Kelvin 'demonstrated' that the Earth's long lifetime estimated by geologists was impossible. Using an apparently rigorous chain of reasoning based on the science of thermodynamics, he 'proved' that, after about 100 million years of existence, the Earth would have lost so much heat that life would have become impossible. Since the Earth had not yet cooled to the point of making life impossible, he reasoned that it must be younger than 100 million years. As is often the case, the reasoning was rigorous but the starting point was erroneous.

The flaw was the implicit assumption, in the calculation of the energy balance, that there was no hidden source of energy inside the Earth.

Since the discovery of natural radioactivity, which provides energy to the Earth, the Earth has been 'given permission' to be much older than 100 million years.

This historical example might encourage those who try to find hidden connections between phenomena that are not yet understood. Scientists may declare that something is 'impossible' just because they don't understand it but, in the case of the memory of water, it could be better to be a little more prudent lest one should look like a fool in a decade or so. It is true that, for the moment, scientists do not have an adequate theory to account for the chemical processes within cellular water or for the effects of magnetic fields on cellular processes. However, when such fields act on cells in a test tube or when they favour certain forms of cancer, they do not ask for permission from the Fellows of the Royal Society or from anybody else. It should be remembered that it is the scientists themselves who must adapt to the natural phenomena and not the other way around. The theory of coherent domains may ultimately prove inadequate to explain such phenomena or may even prove to be wrong, but even if this were so the problems discussed here would not thereby vanish – on the contrary, they would only become harder to solve.

Before concluding this chapter, I wish to stress again that the issues raised involve questions rather than answers. Those who challenge conventional theories of what happens when electromagnetic fields interact with water in living matter may not necessarily be completely consistent in their ideas at the moment. However, they have the ability to acknowledge and examine the contradictions between the old theories and a number of observations, some of which are not new but have been known about for a long time. In other words, I stress the inadequacy of orthodox theory rather than the strength of the tentative new one.

Of course, I am talking here of intellectual strength, not of institutional strength. According to Planck, whose unorthodox ideas were the starting point for quantum mechanics, it is not because the opponents of new ideas convert that these new ideas become accepted but because these opponents gradually die out. I am trying to challenge this pessimistic view with the hope that I may hear the end of the story while I am still alive.

The manner in which the theory of coherent domains has been received so far by the establishment can be summarized by the title of this section: 'I don't want to hear about it'. The article describing an important result of this theory appeared in the top journal of physics (*Physical*

Review Letters) in 1988. Because of the significance of the claims contained in this article, one might have thought that it would soon be widely discussed and quoted, if only to be criticized. However, in the major journal of bibliographical studies (*Science Citation Index*) I found only two references to the article on coherent domains which did not originate from Benveniste's group itself.[2]

Appendix 1 contains another example of the tendency of scientists to shun embarrassing questions: that of an anomaly initially known as 'poly-water'. In this case, most scientists have used the fact that the discoverer of this anomaly made an error in his initial interpretation to dismiss the anomaly altogether. The embarrassing question thus dismissed is: how can pure water dissolve enough glass to produce a silica gel? Another case of dismissal of an anomaly because of a possible error in its initial interpretation is the thermal process initially known as 'cold fusion'.

To conclude this chapter, I would like to examine the idea of scientific 'territories'. Like many who write about science, I sometimes refer to research 'areas' and to 'frontiers' between physics, chemistry and biology. These divisions are justified from a sociological point of view, in that academic disciplines are distinct objective realities, embodied in university departments, learned societies and scientific journals. But they are man-made realities, not natural ones. One should preferably consider physics, chemistry and biology as three aspects of the same natural phenomena – in other words, when atoms within living cells get rearranged because of the influence of oscillating fields, they do not ask themselves if this modification is physical, chemical or biological. The problem is that most scientists are trained within a particular discipline, a process which frequently renders them unable to consider all the various dimensions of a phenomenon. This contributes to their tendency to reject 'borderline' phenomena and adopt an attitude of 'I don't want to hear about it'. The multidisciplinary nature of the memory of water is one feature which it shares with other anomalies mentioned in this chapter. The inability of physicists, of chemists and of biologists to talk to each other or to examine phenomena which lie outside their own field, and of which they have limited expertise, has been one of the reasons why subjects such as the memory of water have been placed in a 'no-scientist's land'. Perhaps UNESCO should set up a department to take care of multidisciplinary phenomena in search of a scientific passport!

By refusing to transcend disciplinary barriers, scientists have been as blind as the characters of a story attributed to Buddha. In this story, a rajah brings four blind men to an elephant. The first one is presented with a tusk, the second one with an ear, the third one with the tail and the last one with a leg. To each one, the rajah says: 'This is the elephant.' He then asks them: 'Have you studied the elephant?' They all answer: 'Yes, your Majesty.' 'So, what are your conclusions?' Each blind man answers in turn: 'The elephant is like a spear.' 'No, the elephant is like a fan.' 'No, it is a sort of rope.' 'No,' says the last one 'the elephant is a kind of tree.' In order to understand the total reality of the elephant, it is necessary to put together the findings from people at different ends of it. The same is true for the memory of water: findings from biology, physics and other disciplines need to be integrated to arrive at the true picture.

CHAPTER TWO

HOMOEOPATHIC DILUTIONS: SHAKE VIGOROUSLY WITHOUT HEATING

THIS chapter is based on a study of Elisabeth Davenas' laboratory books. Between 1985 and 1989, she wrote 2000 pages of notes concerning 500 experiments. In the following pages, I present the results of my investigation with the cautiously optimistic hope that it will eventually lead to a crack in the wall of the scientific dogmatism which has prevented any serious examination of the memory of water.

The tool used to study the biological properties of high dilutions was a standard one: the staining of certain biological cells called basophils, which are a type of white blood cell that plays a role in the organism's immune defence system. The report contained in this chapter is not essentially different from the one by Davenas, Benveniste, Poitevin and 10 other scientists published in *Nature* in June 1988, the main difference being that it is both more detailed and less technical. In order to make reading of this book easier for the non-scientist, most of the technical details have been collected in the Appendixes. The report of the high dilution experiments is in Appendix 2 *(see also Appendix 6d)*. However, this is still the most difficult chapter. If you have difficulties in following some of the technicalities, don't be discouraged; after this chapter, it will get easier.

AN ULTRASENSITIVE BIOLOGICAL TEST: THE STAINING OF BASOPHILS

The instrument employed by the team led by Benveniste was a biological test with which they were familiar because they had developed it some 10 years before. This test had been patented by INSERM, the French National Institute for Health and Medical Research, and concerns the staining properties of basophils. Like many biological cells, basophils have a jelly-like appearance. Their transparency requires them to be stained with some sort of dye, in order to make them visible. Staining is a fundamental biological technique, as shown by the etymology of the word chromosome (it is derived from two Greek words meaning 'colour' (*khrôma*) and 'body' (*soma*)).

The capacity of basophils to absorb the particular dye used in the experiments depends on a large number of factors, including the health of the person whose blood is being tested. In the research on high dilutions, the factor being analysed was the manner in which a certain molecule influences the staining properties of basophils.

aIgE

The molecule in question in the research described here is anti-immunoglobulin E, (also known for short as anti-IgE or aIgE). There are three important things about it.

First it is able to inhibit or even remove the staining of basophils to the point that they sometimes all become invisible, as if they had never been stained. It can be thought of as a biological 'paint stripper' erasing the dye (which renders them visible) from some of the basophils. The percentage of basophils that stay invisiblc naturally depends on the quantity of 'eraser' being used; therefore, the more it is diluted, the less efficient it should become. As we shall see, however, this is not always so; some high dilutions act as 'erasers' even when there is no molecule of 'eraser' left.

The second property of the aIgE molecules is their large size, which makes it possible in the high dilution experiments to distinguish between biological effects due to the actual presence of the molecules and the same effects due to the 'memory' of these molecules stored by water. By use of a standard technique of calibrated filtration to eliminate large molecules from the solutions, it was possible to check that the biological activity observed with some high dilutions of aIgE was not due to residual aIgE molecules that might have escaped elimination during the successive dilutions.

The third property of aIgE is that the 'eraser's' efficiency at the optimal concentration differs greatly from one person's blood to another. This variability of the staining reaction is not a characteristic of high dilutions of aIgE but is due to the reaction itself, whose delicate mechanism can be influenced by a large number of factors which are impossible to control and some of which remain unknown. In some cases, a given amount of aIgE will succeed in preventing the staining of practically all basophils. In others, the same amount of aIgE will have no significant effect on the staining properties of the basophils. The maximum proportion of basophils losing their visibility is usually somewhere between the two extremes, with an average of about 50 per cent.

In order to study the possible action of high dilutions of aIgE in the absence of aIgE molecules, scientists were forced to use a very sensitive test. As is often the case, they had to reach a compromise between opposite requirements of sensitivity and reproductibility. The inhibition of staining constitutes a very sensitive test of the presence of aIgE molecules. However, like most other ultrasensitive instruments this test has a drawback: its results can vary with the slightest change in experimental conditions.

THE SCIENTIFIC STUDY OF HOMOEOPATHIC DILUTIONS

The Experimental Method

The experimental method is probably the most powerful tool of the scientific kit. It attempts to establish a relationship between an observable phenomenon and a 'cause' that can be isolated. Let us begin by considering a simple example of the method and of the difficulties one might encounter in practice. Imagine a Martian who believes that the taste of seawater is due to the presence of small animals. How would you go about convincing him that the taste is actually caused by the dissolved salt?

You might begin by boiling the water in a pan for a few minutes to kill the 'animals'. After it has cooled, you would ask the Martian to taste the water and see for himself that the taste was still present. If he then told you that his animals had survived, you might respond by boiling the seawater more vigorously until all the liquid had evaporated, leaving only salt at the bottom of the pan. You could then add this salt to a glass of tap water and ask him to taste it. 'I can't taste anything,' he might say, 'the "spirit of taste" left the water when you boiled it.' He is thus apparently incapable of tasting the salt when he believes that there is nothing to be tasted. So, in order to convince him, you might try the following 'blind' experiment. You first boil a large quantity of seawater to evaporate it, collecting the salt that remains and also condensing the vapour that comes off to collect pure water. You then fill 10 small glasses with this water and tell your Martian: 'Now I am going to leave the room. You will pour one spoonful of the powder in one of the 10 glasses and stir it thoroughly. In 5 minutes, I shall come back and test each glass.' After having tasted the water, you tell him correctly which of the 10 glasses contains the salt. 'Yes,' he might say, 'you guessed the right glass, but you were just lucky.'

If you still have some patience left, you might propose repeating the experiment 10 or 20 times. Ideally, after performing this experiment sufficient times the Martian should agree that your idea was correct. However, even if he never made any error in the coding and even if you succeeded in designating the correct glass most of the time, it is still possible that he might tell you: 'I don't know how you did it, but there must be some trick. I *know* that the taste of seawater comes from an animal living in the sea.'

The reader will have to wait for the second part of the book to appreciate fully the relevance of the above sentence to the 'memory of water' experiments, and the resources available to those who refuse to believe something that they don't understand. Actually, scepticism about experimental evidence is often justified, as no experiment, however well conceived and well performed, can provide absolute proof. The only science in which such proof is possible is mathematics. In the other sciences, however, one can produce only evidence. Part Two of this book is devoted to the question of evidence and to the attitudes of people in the face of it concerning the memory of water. For the moment, I simply want to point out the two basic elements of the experimental method.

1. Two situations are created which are identical in all respects, except for one: the factor to be studied. In one situation, called the control situation (C), this factor is missing; in the other, the experimental situation (X), it is present. (In the above example, the factor that varied was the presence or absence of powdered salt.)
2. If, in a sufficient number of experiments, the outcome or result (in the example the taste of the water) is significantly different in situation X to that in situation C, then this difference is assumed to be due to the presence or absence of the factor concerned.

The Experimental Method in High Dilution Experiments

When studying high dilutions, the best experimental design is the one represented in Figure 2.1. In the first phase, two high dilutions are prepared in exactly the same way. The only factor that changes is that, in one case, the substance in test tube 1 is X (for instance a solution of aIgE in water) and in the other it is C, the control dummy product (for instance, the same water as used to dissolve aIgE). The chemical X is biologically active (i.e. it 'erases' the dye of a significant percentage of stained basophils) whereas the control C is taken to be inert (i.e. having no 'erasing' proper-

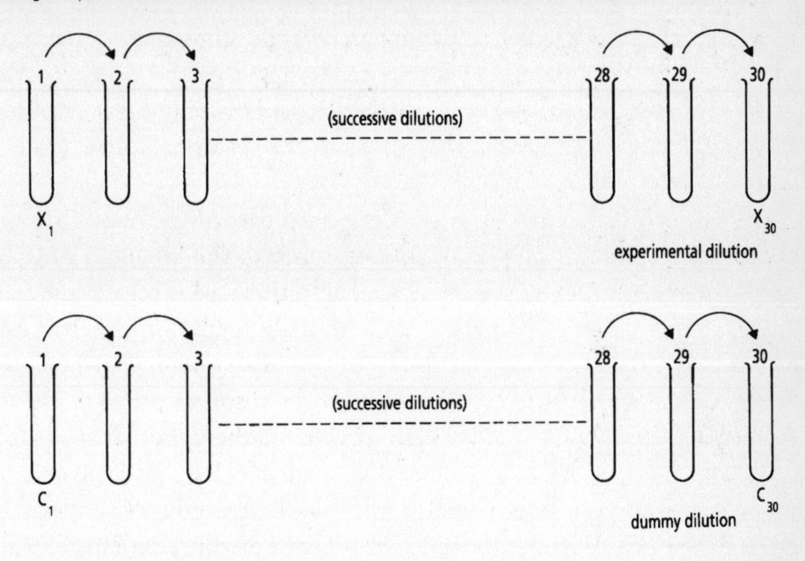

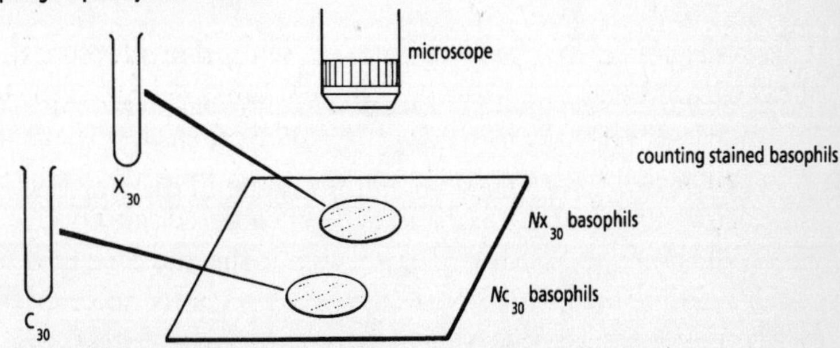

FIGURE 2.1 Comparing a homoeopathic dilution to a dummy one

ties). Each of the initial products undergoes a certain number of divisions by 10 followed by vigorous shaking. The example in Figure 2.1 shows the preparation of the 30th decimal dilution of both X and C.

In the second phase of the experiment, the biological potencies of the high dilutions prepared in the first phase are tested; the high dilutions are added to basophils which are then counted under the microscope. The result of the test of high dilution (X_{30}) is said to be positive if the number of basophils which are visible under the microscope is lower than that for the corresponding dummy high dilution (C_{30}).

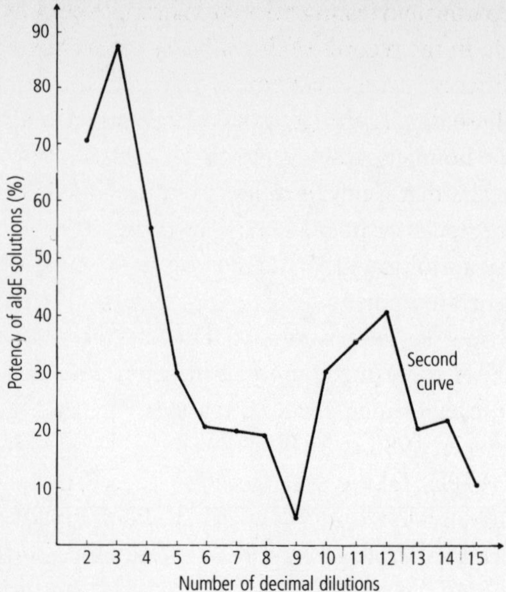

FIGURE 2.2 The 'second curve'

EVIDENCE FOR THE POTENCY OF SOME HOMOEOPATHIC DILUTIONS

In 1983, Bernard Poitevin (a research worker interested in homoeopathy) contacted Benveniste with a research proposal concerning the biological effects of some homoeopathic dilutions. Benveniste expressed his scepticism but accepted the proposal. I was not there to witness his initial reserve, but I found material evidence of this scepticism in Elisabeth Davenas' laboratory notebooks.

On 5 November 1985, Elisabeth Davenas and Francis Beauvais observed for the first time the quite unexpected phenomenon which is illustrated in Figure 2.2 and which they later called 'the second curve'. The potency of the biological 'eraser' (aIgE) is at a maximum at a concentration corresponding to the third decimal dilution and, as expected, then decreases with each successive dilution. However, after the ninth dilution the situation changes: the solution's potency *reappears and starts to increase*, in spite of the fact that the *quantity* of aIgE is still being divided by 10 at each successive dilution.

Ten days after this discovery, the scientists involved initiated a series of 'blind' experiments using three different homoeopathic solutions to inhibit the action of aIgE. In these experiments, Elisabeth Davenas was given two identical tubes labelled only 'A' and 'B' or '1' and '2', one of the tubes of each pair containing a solution of the inhibiting product and the other only water, hence when preparing and testing the high dilutions she could not be aware which was which. In the records of the first 6 months of experimentation, I found that 12 different people had been involved in the coding of 34 blind experiments. These details are mentioned to illustrate Benveniste's initial scepticism about homoeopathic dilutions in contradiction to the claims from some quarters that Benveniste had acted as a credulous scientist, by accepting bizarre results without adequate evidence.

Following the inhibition experiments just mentioned, these INSERM scientists concentrated on the direct effect of high dilutions of aIgE on the staining properties of basophils. (The interested reader will find details about these experiments and about their results in Appendix 2. Critiques of these experiments will be analysed in Part Two, especially in Chapter 6.) Between 1986 and 1990, some 250 direct experiments were performed with aIgE. Taking everything into account, the results represented impressive evidence in favour of the memory of water. In this section, however, I will only present one experiment in order to illustrate the principle behind the various tests.

In this experiment, a series of 25 decimal dilutions of aIgE was compared with a corresponding series of 25 decimal dilutions of the dummy product. The range of dilutions was quite large, from 21 to 45. A

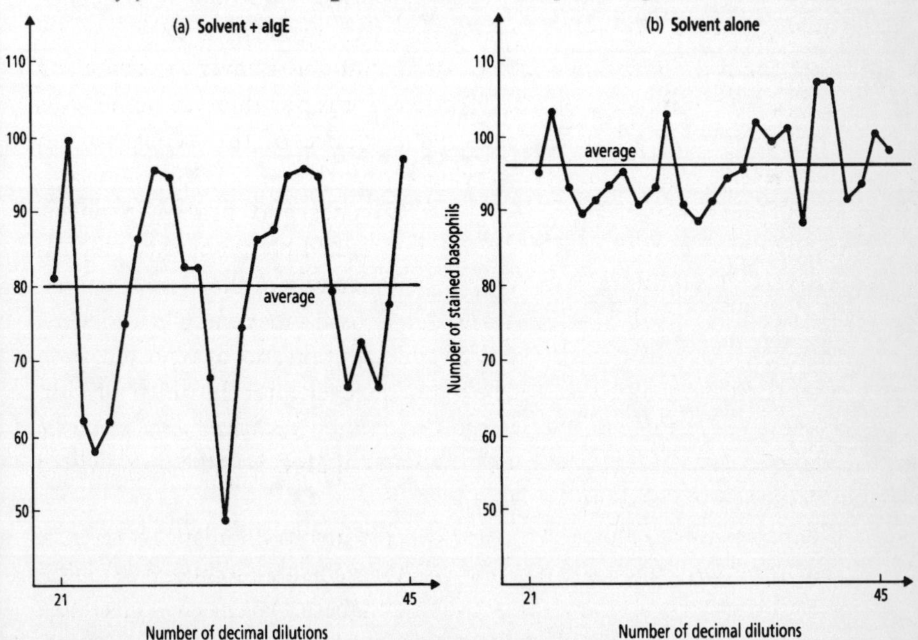

FIGURE 2.3 Results of a basophil-staining experiment

graph of the results is shown in Figure 2.3. Graph (a) shows the number of basophils which remained stained at successive high dilutions of aIgE solution, while graph (b) shows the corresponding numbers of basophils with successive dilutions of the dummy solution. This experiment indicated that the memory of water manifests itself in three different ways:

First effect (test 1): on average, the number of basophils that remain stained is *lower* when they have been impregnated with high dilutions of aIgE than with high dilutions of the dummy solution (80 instead of 96).

Second effect (test 2): the number of stained basophils is more *variable* when they have been impregnated with high dilutions of aIgE than with high dilutions of the control.

Third effect (test 3): the left-hand graph (a) seems to oscillate as a periodic wave.

From a statistical point of view, each of the three effects outlined above provides an independent test of the memory of water. A fourth type of test described in Appendix 2 reveals an effect due to heat: when high dilutions of histamine were heated, they lost their ability to inhibit the effect of aIgE, *even though the original histamine molecules are not themselves sensitive to heat at the temperature used.*

QUANTITATIVE SUMMARY OF RESULTS OF HIGH DILUTION EXPERIMENTS

Putting together all the experiments which fulfilled the stringent conditions described in Appendix 2, I obtained the results shown in Table A2.3. The probability of getting such results purely by chance is negligible. This probability is the same as that of coming up tails 326 times when tossing a coin 453 times.

Each series of results by itself is so far from the result you would expect by chance that it is very significant from a statistical point of view. Taken together, the results become literally impossible to explain through chance alone. To understand this, imagine a massive team of scientists capable of performing all the experiments summarized in Table A2.3 in a single day (instead of in 4 years as the small INSERM team did). In order to reproduce a pattern of results as significant as those shown above by chance alone, they would have to repeat the experiments for billions of years! So these results cannot be disposed of by invoking statistics; if they should turn out to be erroneous (and this can never be excluded in any experiment), the source of the error would have to be sought elsewhere.

THE SCIENTIFIC STATUS OF HIGH DILUTION EXPERIMENTS

I would like to conclude this chapter with a table summarizing the high dilution experiments which have been performed with basophils by Benveniste's group (Table 2.1). There are two reasons for including it. First, it clearly shows the large number and the variety of experiments that have been performed. Secondly, it illustrates the major theme of this book, which is censorship within science and suppression of available evidence. Comparison between the two right-hand columns in the table

TABLE 2.1 500 *high dilution experiments using basophils*

Type of research	*Number of experiments* *Before* *the inquiry by*	Nature *After*
Evidence for high dilution effects		
Direct effects (aIgE)	108	128
Inhibition (Apis)	41	65
Other chemical		
PLA$_2$ (pig)	10	10
PLA$_2$ (Naja naja)	4	
Histamine	48	
Lung histamine	22	
Apium virus	4	
BOB	3	
Calcium ionophore	5	1
PLC	1	3
Mellitine	4	
Successive dilutions		
Waves[a]	10	
Dilution after 100th	8	
Position on the 'wave'	5	
Pools of dilutions	4	
Influence of other physical factors		
Temperature	25	
Shaking	8	
Calibrated filters	13	
Solvents other than water	13	
Dilutions by nine	3	
Direct current	1	
Alternating current	1	
Magnetization		9
Transmission	1	
Drying	3	

[a] At least 20 successive dilutions after the 18th decimal dilution.

suggests that outside pressures effectively stopped ongoing research. The second column indicates the number of experiments of a particular type performed before the investigation by *Nature*, whilst the third column shows experiments performed afterwards. The relative variety of the research carried out before *Nature*'s inquiry is clearly visible.

A range of different types of exploratory research were also performed. For example, in 10 experiments the forms of the periodic waves were studied with a large number of sequential high dilutions. In eight other experiments, this effect was examined with unusually high dilutions, going beyond the 100th decimal dilution. The previously mentioned destructive effect of heat on the potency of high dilutions was systematically studied in 25 experiments. Eight experiments looked specifically at the influence of vigorous shaking. Altogether, 14 different types of experiments were performed to investigate the effect of physical parameters. The numbers in the third column demonstrate that, after the visit of the fraud squad sent by *Nature*, this innovative research was essentially stopped. Ninety-five per cent of the 200 experiments performed after that visit were a sheer repetition of previous experiments. For 2 years, scientists essentially repeated the same two experiments concerning the direct effect of high dilutions of aIgE on the staining of basophils and its inhibition by high dilutions of a product called Apis mellifica (an extract of bee venom).

In this chapter, I have described the most detailed laboratory study of homoeopathic dilution published so far. What makes this particular study so special is the fact that it is the only one where a systematic investigation of physical conditions that might influence the 'memory of water' was initiated, in order to gather facts that might finally lead to a full understanding of the mechanisms behind that memory. This may explain the violence of the attacks against the basophil experiments; these will be described and analysed in Part Two.

Despite its scientific importance, it should be stressed that the basophil study directed by Benveniste is far from being the only scientific study of extreme dilutions. Even excluding publications by journals devoted to the scientific study of homoeopathy (and this of course is a form of censorship), I found 25 scientific articles published by 17 different groups of scientists reporting high dilution effects (*see Appendix 6d*). So far, the scientific community has succeeded in maintaining the memory of water within the ghetto of homoeopathic research. It will be interesting to see whether a debate about the memory of water eventually occurs.

CHAPTER THREE

AGENT X TRAVELS THROUGH WALLS

TRANSMISSION EXPERIMENTS DISMISSED AS 'BLACK MAGIC'

We are all familiar with stories where a police inspector is faced with the following 'impossible' situation: a corpse has been found in a one-room apartment with no windows and the door locked from the inside. In spite of a very thorough search, no weapon could be found in the room, and of course no murderer. According to the medical examiner, death (caused by a bullet wound) was instantaneous so that the victim could not possibly have locked the door after having been shot. In this kind of fiction, one often finds two types of investigators; one is dumbfounded and keeps repeating 'Fantastic!', while the other is busy thinking of various possible explanations to the crime. The first holds firmly to the idea that the crime could not possibly have occurred; the other one proceeds from the only thing that he is sure of, namely the presence of the corpse lying on the floor.

In the fictitious situation described above, the enigma is created by three apparently contradictory assumptions: the first is that the report of the policemen who broke into the room is reliable, at least as far as the locking of the door is concerned; the second is that the medical examiner's report is reliable as to the cause and speed of death, which excludes the possibility of the dead man having locked himself in; the third is that the murderer and the weapon were not hidden in the room at the time that the policemen broke in.

After having convinced himself of the validity of the above statements, the intelligent inspector might ask himself: 'If the door cannot possibly have been locked from the inside, could it have been locked from the outside?' Through a careful inspection of the lock, he might then find that it could indeed have been operated from the outside by radio control.

Apart from their potential significance for science and technology, what I find fascinating in transmission experiments is that they represent a similar kind of logical enigma. As we shall see, the experimental design is so simple that it appears to have few possible loopholes. I myself have been unable to think of one and I am not aware of anybody else who has. The alternative 'explanations' that I have heard of so far are simply denials that the experiments could possibly be genuine; one scientist talked of the possibility of fraud, while another described Benveniste's experiments as 'black magic'.

For the moment, the bare outline of one of the early transmission experiments will suffice to illustrate the analogy with the above fictitious situation. A glass sealed phial containing the active agent was placed on a coil at one end of an electrical transmission machine and a sealed phial containing pure water placed on a second coil at the other. The machine was turned on, and the water in the sealed phials was 'treated' for 15 minutes. This water when tested was found to be biologically potent, while that in untreated 'control' phials was inactive. It appeared that some active agent had travelled from one sealed phial to the other, through both the glass walls of the tubes and the apparatus.

THE EARLY HISTORY OF TRANSMISSION EXPERIMENTS

In the previous chapter, I mentioned that Benveniste's research on the memory of water was stopped after the visit of *Nature*'s self-appointed experts. Table 2.1 shows that one of the lines of research explored concerned these transmission experiments. In June 1988, a homoeopathic doctor called Attias convinced Benveniste to try out an electrical machine which he claimed transmitted chemical information. At the time, Benveniste had just learned of the existence of the theory of Preparata and Del Giudice. In his paper published by *Nature* on high dilution experiments, he mentioned the possible part played by electrical or magnetic fields in the memory of water, but made no reference to the theory of coherent domains.

It is difficult to reconstruct these events so long after the facts, but I imagine that, since Benveniste was at a loss to understand the potency of homoeopathic dilutions which he had been studying for several years, he must have thought: 'Why not give it a try?' Whatever his reasons, a transmission experiment using the electrical machine brought in by Dr Attias was subsequently performed. The result recorded by Elisabeth Davenas in her laboratory book was positive, and she expressed her perplexity in this report.

Two weeks later, the fraud squad sent by *Nature* spent 5 days in Benveniste's laboratory. (This visit will be described in detail in Chapter

6.) A report entitled 'High dilution experiments a delusion' was published by *Nature* a few weeks later. Fraud was not directly mentioned in this official report, but the visit had both a traumatic effect on researchers studying homoeopathic dilutions and a devastating effect on the public image of their research.

At the time of my first contact with Benveniste in March 1992, he was studying high dilutions with a system that had nothing to do with the staining of basophils. It was a sensitive biological system that has been used to test new drugs in pharmacology for about a century, in which the key part is the heart of a guinea pig (or rat) which has been appropriately immunized in order to render it especially sensitive to the substance being tested. Benveniste was, in addition, studying homoeopathic dilutions of different substances; the one most often used was ovalbumin, a protein contained in egg white. The only resemblance of the new system to that using basophils was its great sensitivity and its correspondingly high variability.

Using this highly sensitive biological detector, Benveniste was able to confirm the potency of some high dilutions. The experimental design was similar to that described in the previous chapter; that is, the effect of high dilutions of an active product was compared with that of an inactive one prepared with the solvent alone.

Some experiments on physical properties of the memory of water were also performed with the new system. The fact that heat destroys the potency of high dilutions was confirmed. With the help of two physicists from a nearby laboratory, a new way of erasing the memory was also found – a low frequency alternating magnetic field. The potency of the dilutions disappeared after the physicists had submitted them to this field, although it remained intact in untreated dilutions which had travelled back and forth between the two laboratories but had not been exposed to the field.

The experiments using magnetic fields were successfully repeated in a blind fashion, in which the physicists coded the tubes with random numbers; the potency of these tubes with unknown contents was then tested on the animal hearts. It was the success of these magnetic experiments that convinced Benveniste that the memory of water had something to do with electromagnetic fields, which probably made the study more attractive to him from a theoretical point of view.

Another reason for attempting these experiments was criticisms of the former high dilution experiments, in which sceptics resorted to making complex assumptions in order to avoid considering the abhorred hypothesis of the memory of water. In the transmission experiments, these objections clearly become irrelevant; in fact, the experimental design of transmission experiments was so nearly ideal that it is difficult to think of any scientific explanation other than that there is indeed chemical information transmitted to the second phial electrically by the machine without any corresponding molecular transport.

Three and a half years elapsed before Attias succeeded in convincing Benveniste to try transmission experiments again, this time in a more systematic manner. I remember a telephone conversation during which Benveniste mentioned to me the possibility of transmitting chemical information with an electrical machine. I was as sceptical as those who first heard about an apparatus supposed to transmit human voices through electrical cables. I consequently became a witness to some of the early trials in which the machine was operated by Attias. After a few trials with this machine, Benveniste had another one built, which he used for his later experiments. This second machine was essentially a low frequency high gain amplifier.

AN OBSERVER BECOMES A PARTICIPANT

As mentioned in the introduction, I had been working for a long time on issues related to the nature of scientific knowledge and was interested both in the communication problems between scientists having different or antagonistic views and in scientists' resistance to innovative research. When I first became involved with the problem of the memory of water, I viewed the polemics about high dilution effects as simply another interesting opportunity to analyse these scientific communication problems.

My attitude was significantly modified, however, in June 1992 during one of the early transmission experiments. The experiment performed that day in front of several witnesses had given intriguing results, but these were not clear cut. One of the other observers was a physicist who was much more competent than I, at least in the matter of the electrical apparatus being used. However, in this instance, it was evident that his technical competence was counterproductive as, instead of concentrating his attention on observing the 'facts', he became stuck in the problem of trying to understand how the electrical machine could possibly produce such effects, and consequently developed a resistant attitude. He was lucid and honest about his prejudices, however, and later exclaimed: 'Don't talk to me about transmission experiments; I have a mental block.' In the discussion that followed, he described the electronic amplifier as 'an infernal machine', and also recalled the scandal created in France by 'sniffing planes'. (These planes were supposed to be able to detect oil but were in fact using a technological trick.)

During my investigation of Benveniste's research on the memory of water, I have had other opportunities to hear people referring to affairs in which fraud or hoax had played a crucial part. When confronted with unexplained facts, most scientists seem unable to remember previous historical examples in which such facts eventually turned out to be the starting point of important new theories; in their selective amnesia they remember only past examples of scientific affairs where proposed heresies were later shown to be unfounded.

I was particularly surprised by the attitude of the physicist above because he of all people had good reason to trust Benveniste's research; he was one of the two scientists who had proposed and carried out the experiments demonstrating the magnetic erasure of the memory of water. During the discussion on transmission experiments, it was he who had confirmed that Benveniste had designated the correct tube as active or control dozens of times without ever making a mistake. His testimony undermined my own scepticism at the time about high dilution experiments and was instrumental in stimulating my curiosity and pushing me into the role of a participant-observer.

A few weeks after those experiments, Benveniste jokingly said to one of his co-workers: 'I bet Schiff is going to ask me for a lab coat and a technician.' I did not ask for a lab coat and I did not increase my visits to Clamart, but I did become personally involved in his transmission experiments. I was beginning to see what was going on as an intellectual and a psychological challenge. I feel that I was in a better position than the physicist above to consider the transmission experiments objectively because I chose to employ a technique of 'black boxing' (*see page 40*) which helped me to suspend judgement and avoid any preconceived ideas about the functioning of the 'infernal machine'. I also had no opinion on the exact nature of what needs to be transmitted in order to modify the functioning of the heart of an animal. My viewpoint was therefore not that of a physicist, of a chemist or of a biologist, but simply of an interested scientist unattached to any given discipline.

A SCIENTIFIC EXPLORATION GETS PARALYSED BY THE BURDEN OF PROOF

Until June 1992, transmission experiments had been very tentative, in particular because of contamination problems in which samples of water that were supposed to be chemically pure occasionally had an effect on the hearts. As we shall see in the next chapter, this eventually turned out to be an interesting phenomenon in itself. In the transmission experiments some of these contamination problems were solved by leaving the samples of water for an hour or two in an incubator at a temperature of 70 °C.

After the preliminary experiments in spring 1992, Benveniste started presenting transmission experiments to scientists who did not belong to his laboratory. The report on the first of a series of seven blind experiments performed in public is reproduced in Appendix 5, together with the reaction of the head of INSERM, who had received a copy of this report. In a letter to Benveniste, he mentions the 'pernicious character of such "information"' and continues 'Should you persist in this type of behaviour, I would be forced to draw serious consequences.' As we shall see in Part Two, INSERM did not simply issue threats, but also took action by depriving Benveniste of his research tool. The threat just mentioned came after a protracted battle about homoeopathic dilutions. It seems that it pushed Benveniste into repeating demonstrations instead of pursuing his research; in other words, the logic of convincing others subverted that of scientific exploration.

The following year, in spring 1993, INSERM announced that experts were going to visit Benveniste's laboratory and would then write a report about his transmission experiments. I feared that, under the guise of a scientific report, these experts would behave in the same manner as had the 'scientific commando unit' sent by *Nature* 5 years before, and considered possible strategies which would bypass official experts. One possibility I entertained was to place an advertisement such as the following in the French equivalent of the *Scientific American*:

HOW MANY SCIENTISTS ARE STILL INTERESTED IN SCIENCE?

Scientificus curiosis is a species that is rapidly disappearing. As members of this species, we are looking for other inquisitive scientists willing to play the role of scientific bailiffs for experiments related to the memory of water. In these experiments, a low frequency amplifier appears to transmit chemical information to pure water, without any molecules being transported. If you wish to receive the experimental protocol, write to:

SCIENCE INNOVANTE,
6 RUE MONTMARTRE, PARIS.

Scientists assured of their knowledge not welcome

At the time, however, Benveniste was fighting for professional survival and trying to maintain his research team within the National Institute for Health and Medical Research: since the above text would probably have appeared as gratuitously provocative I refrained from presenting this suggestion. It is impossible to know what would have happened if Benveniste had fought his battles on the front of academic freedom, without being so keen on convincing his colleagues of the reality of his observations. Instead of putting all his efforts into the formidable task of convincing those who refused to be convinced, he could have continued his exploratory research. However, the design of the transmission experiments was so simple, and in principle so convincing, that Benveniste hoped that, by repeating and improving this basic experiment, he would finally convince leaders of the scientific establishment. In spite of their basic simplicity, Benveniste's demonstrations have not yet achieved this goal.

I believe that, instead of hoping for a recognition that was denied *as a matter of principle*, without any serious technical examination of the case, it would have been wiser to look for allies outside the narrow circle of established scientists. Scientists working in marginal fields should refuse to let a debate be restricted to its apparently 'technical' aspects. Having said this, it must be admitted that my position was more comfortable than Benveniste's; I had nothing to lose, whereas he was being threatened with a definitive closing of his laboratory.

More generally, we also had different views about the significance of 'crucial experiments', where Nature is supposed to speak in an unambiguous manner. Most scientists believe in the possibility of such experiments, and Benveniste shared this belief with his opponents while I did not, at least from a practical point of view. From the latter point of view, presenting a good scientific case is only one part of the problem. One must also have genuine discussions with people who are both knowledgeable and willing to learn something new, and this may be very difficult to achieve.

DESCRIPTION OF TRANSMISSION EXPERIMENTS

To bring home the point that the following report on transmission experiments has, like the rest of this book, been my sole responsibility, I have concentrated my analysis on those experiments in which I have personally participated. In the description that follows, I will adopt the point of view of an outside observer of these experiments and in addition use the idea of 'black boxing', an idea which has helped me to concentrate on the enigma raised by transmission experiments without getting lost in a maze of irrelevant details (*see next section*).

From the point of view of an outside observer, the transmission experiments can be described in the manner represented in Figure 3.1. The transmission machine is composed of two coils that are literally 'black boxes' connected by an electric wire. After switching the power on, the operator places a tube containing ovalbumin (the white of an egg), marked 'OVA', on the input coil of the machine and an unmarked one on the output coil; both tubes contain transparent liquid.[1] After leaving the

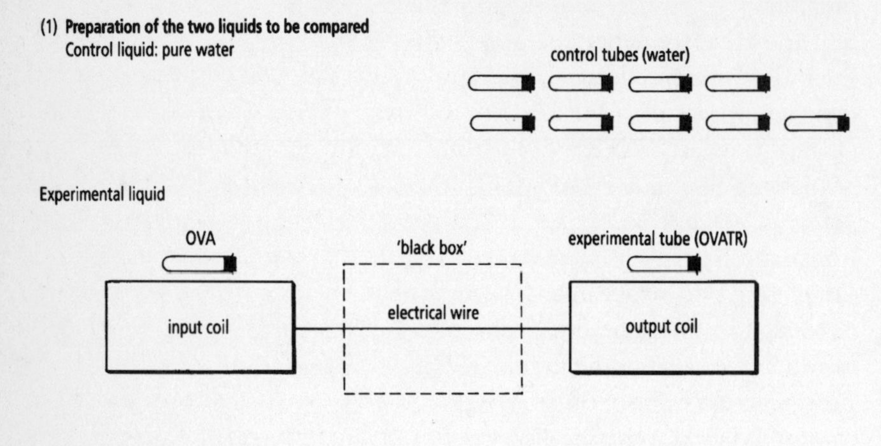

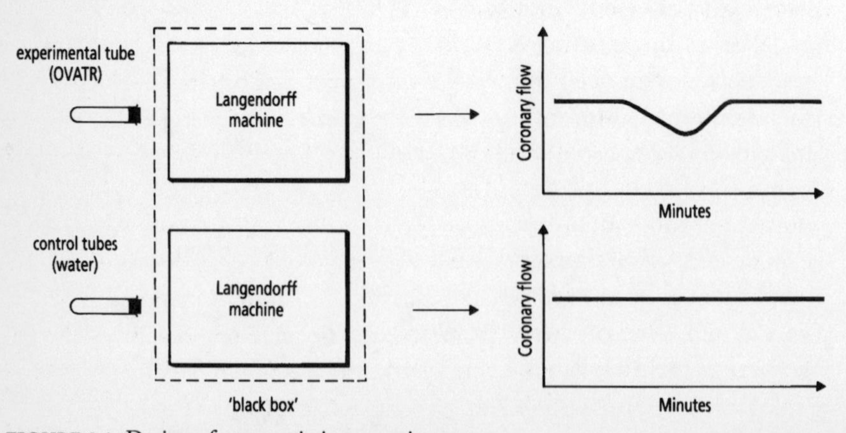

FIGURE 3.1 Design of a transmission experiment

output tube on the machine for 15 minutes, the operator removes it and labels it 'OVATR' (transmitted ovalbumin). Using a syringe, he then injects some of its contents into the biological testing system (called a Langendorff apparatus), which contains the heart of a freshly killed guinea pig or rat. Underneath are graduated tubes which receive liquid dripping from the Langendorff apparatus and which are changed every minute by a rotating machine. Every minute, the operator takes the last graduated tube, reads the level of the liquid and writes it down on a sheet of paper. The quantity of liquid that has dripped into the graduated tube is a measure of the coronary flow of the heart used to test the various liquids.

The measurements collected by the operator appear as a series of numbers presented as a table, each column of which corresponds to a particular liquid. One column of numbers represents 'Ovalbumin transmitted' (OVATR). Supposing that, in this column, you see the following figures corresponding to the levels of liquid read by the operator: 4.2, 4.3, 4.2, 3.7, 3.4, 3.9, 4.1, 4.2, 4.2. This would mean that, 4 minutes after being injected, the heart flow dropped by 20 per cent and then returned to its initial value. In the other columns, which corresponded to the same liquid untreated by the transmission machine, no drop in the value of the numbers was observed, therefore no variation of the heart flow occurred. I might tell you that you had just witnessed the transmission of a chemical agent without matter being transported. 'But,' you would probably say, 'I have seen nothing of the sort! All I have seen are differently marked tubes which have been moved around, followed by columns of numbers on a sheet of paper.' You are perfectly right.

In order to be convinced that you really have witnessed something as puzzling as the transmission of a chemical agent without matter being transported, you need to concentrate your attention on the first part of the experiment, where the tubes are placed on coils at each end of the machine. The best way to do this if you want to be sure is to perform this part by yourself and then give the Langendorff apparatus operator a number of tubes identified by numbers only. This would then be a 'blind' experiment. The simplest blind experiment which you might want to perform could be as follows. The operator gives you a set of tubes filled with liquid. One of the tubes is marked 'OVA', while all the others are unmarked identical tubes. Once you are alone, you place the tube marked 'OVA' on the input coil of the switched-on machine and you choose one of the unmarked tubes, which you place on the output coil. After leaving the tubes in place for 15 minutes, you write a number between 1 and 10 on the unmarked tube. You then take nine other unmarked tubes and give them different numbers. You finally give the 10 coded tubes to the operator (one experimental tube with 'transmitted ovalbumin' and nine dummy ones).

If the operator correctly identified the experimental tube, you would begin to think that there was indeed something interesting going on as there is only a 1 in 10 chance of the operator being right. Of course, to be more certain, you would have to repeat this experiment several times and, in order to avoid possible bias, each trial should be performed in such a manner that the operator does not know in advance the identity of the tube whose contents he is testing. This is what I did, as I explain in the next section. But first let us clarify the idea of 'black boxing'.

Black Boxing

Someone interested in the mystery of the transmission machine can start with asking the following question: 'In order to look and see whether something unusual is indeed happening, exactly what do I have to know about the experiments and what can I leave out?' I call the parts I can leave out at present 'black boxes', because it is not essential for me to check and to understand exactly what is going on in this part of the experimental setup in order to decide whether the phenomenon has any significance. A black box, therefore, designates an indeterminate part of the setup, or connection between parts, whose details are considered of no immediate importance.

Several types of black box are used in the transmission experiments. The first is a physical one – the transmission machine itself. All one needs to know about it is that it has an input where an active tube is placed and an output where the tubes of water are deposited for 15 minutes. The second black box is a biological one containing animals that have been immunized in the proper way at the right time, as well as the Langendorff system (including the live heart of an animal, plastic tubes with liquid going through the heart, graduated tubes to measure the heart flow, etc.). As shown in Figure 3.1, the input to this black box is a test tube, while the output is a line on a graph indicating how the heart flow was modified by the content of this tube. Thirdly, we have two chemical black boxes – one the tube marked OVA, which is supposed to contain a solution of ovalbumin, and the other the series of unmarked tubes supposed to contain pure water.

In order to illustrate the fact that a critical observer does not need to have the same inside knowledge of the black boxes as do the scientists who use them, let us consider the second of the chemical black boxes – the set of unmarked tubes. The experimenter must make sure that the water in the tubes is biologically pure, and when one uses an ultrasensitive detector this is no small achievement. If the water is contaminated, it will be impossible to distinguish the tube of 'transmitted ovalbumin' from the dummy ones; this possibility of contamination of the dummies can render some experiments insensitive, so that experiments have to be repeated many times to achieve statistical significance. (Problems with the dummy solutions were in practice fairly frequent, especially in the beginning.) But this is the experimenter's problem; it does not directly concern the critical observer. The latter, in contrast, needs be sure of only one thing: the fact that *before the transmission experiment, all unmarked tubes were originally equivalent* (statistically speaking, they were 'part of the same population'). In other words, the difference between 'activated water' ('transmitted ovalbumin') and the other tubes must be unambiguously ascribable to the operation of placing a tube on the output of the switched-on machine for 15 minutes while a tube marked OVA is lying on the input. Therefore the technical problems of the experiments can most usefully remain as a black box for the observer, who can then concentrate on determining whether or not a relationship exists.

In short, what I have witnessed leads me to the following conclusion: some change that is detected biologically can be statistically associated with the conjunction of two events, that is, (1) the electrical machine has been turned on, and (2) a tube filled with liquid has been lying on the input part of the machine.

A PERSONAL TESTIMONY CONCERNING 10 TRANSMISSION EXPERIMENTS

I personally watched all blind transmission experiments, except for the first, that were performed publicly between July 1992 and December 1993. As a matter of fact, I organized and supervised four of these. At the end of 1993, I also performed three experiments by myself, without any other witness. All public experiments were performed in such a way that nobody, including the witnesses, could possibly know the correspondence between the code numbers and the identity of each tube before the code was broken. The system of coding used to achieve this goal is described in Appendix 3a. You would have to assume a very elaborate system of cheating, involving many scientists (including myself), in order to claim that Benveniste or his co-workers used some means of identifying the 'transmitted ovalbumin' other than the one reported in their experiments.

The conclusion that Benveniste really did observe an important phenomenon seems to me difficult to avoid. 'Something' travelled from a closed tube to another closed tube, via an electrical device; that 'something', whatever it may be, was then detected in apparently pure water. The results of the 10 blind experiments analysed in Appendix 3a are presented in Table 3.1.

TABLE 3.1 *Results of 10 blind transmission experiments*

Date	*Device used*		*Mean drop in heart flow (%)*			*Differences observed*
	No. of tubes	No. of hearts	OVA (a)	OVATR (b)	Water (c)	((b-c)/a)
9/7/92	10	2	100	58	5	+0.53
28/9/92	15	7	25	14	6	+0.32
21/4/93	3	4	37	38	4	+0.92
13/5/93	10	2	73	51	4	+0.64
13/5/93	10	2	39	3	4	-0.03
13/5/93	10	2	38	2	4	-0.05
13/5/93	10	2	33	14	2	+0.36
8/12/93	10	2	61	15	6	+0.15
29/12/93	8	2	15	9	4	+0.33
30/12/93	8	2	26	12	3	+0.35

The first column of the table gives the date when the transmission experiment was performed. The biological analysis was usually started on the same day and lasted a few days, depending on the number of tubes that were being tested (second column) and the number of hearts involved (third column).

The next three columns show the 'raw' percentage variations in heart flow induced by ovalbumin, 'transmitted ovalbumin' and pure water respectively. The column headed OVA can be taken as a standard for calibration, as the numbers in this column indicate the percentage variation observed in the heart flow when the hearts were stimulated by a standard solution of ovalbumin. However, it can be seen that the sensitivity of the measuring system was highly variable (as with the basophils), ranging from a maximum of 100 per cent in July 1992 to a minimum of 15 per cent on 29 December 1993. The final results of each experiment (last column) were therefore obtained by comparing the *difference* between the effect observed with 'activated water' (OVATR) and that obtained with untreated water, and then *dividing* by the calibration percentage. For the first experiment, the difference was 53 per cent, which when divided by the calibration percentage (100 per cent) becomes + 0.53 (and so on).

When considered as a whole, the above results are statistically significant. The calculations outlined in Appendix 3a show that the probability of obtaining such results by chance alone is less than 1 in 1000, even when the analysis is restricted to the 10 experiments presented here. In summary, I am not saying that I am certain that Benveniste is right when he claims that the dozens of experiments which he has performed demonstrate the existence of an electromagnetic transfer of chemical information; what I am saying, however, is that the design of the experiments above is sufficiently sound to eliminate other 'confounding' explanations such as:

1. chemical or biological contamination in the test tubes;
2. random errors, in particular those due to an erratic behaviour of the biological measuring system.

In 1994, Yolène Thomas, a biologist working with Benveniste, started to use a new, cellular system to study transmissions. This gave spectacular results, which I was so impressed with that I gave up another project in order to work part-time with Thomas on her experiments. These later results are presented in Appendix 3b. Here I will mention only two. Out of 20 blind comparisons made with coded tubes, in 19 the correct tube was selected. Even more impressive, one prediction of Benveniste's hypothesis (electromagnetic transfer of molecular signals) seemed to have been fulfilled: the effect of the transfer disappeared in output tubes shielded with an alloy designed to stop magnetic fields, while it could be observed in similar unshielded tubes placed concurrently on the same output coil. It should be noted that the transfer observed by Thomas was not a two-step process, as is that displayed in Figure 3.1 (where water acts as an intermediary recipient of whatever is emitted by the input tube); in these cellular experiments, the mysterious 'X' acts *directly* on the cells.

THE SIGNIFICANCE OF TRANSMISSION EXPERIMENTS

As shown in Appendixes 3a and 3b, the bulk of the effects observed can hardly be attributed to statistical artefacts. The fact that transmission effects have now been observed with a biological system very different from that originally used by Benveniste provides an argument against the idea that his transmission results were simply due to some biological artefact of measurements made on hearts. However, in order to eliminate finally the possibility of an artefact (whether biological, chemical or physical), it will be necessary to find some theoretical explanation for the transmission effects observed; in this respect, the 'black boxing' used in my presentation of transmission experiments is insufficient. However, the detailed study of the phenomenon described in this chapter will require the collaborative efforts of several laboratories and several disciplines. A prerequisite for such an effort is an open attitude, admitting the possibility that the study of transmission phenomena might lead to a breakthrough in biochemistry and in biophysics.

Like high dilution experiments, transmission experiments using water as an intermediary support for transmitted molecular signals present a theoretical puzzle: how can information be stored by liquid water without being immediately destroyed by thermal agitation? The theory of coherent domains outlined in the first chapter might provide an explanation for this puzzle. Whatever the explanation, the fact that both types of activated water (from high dilutions and from transfer) lose their potency around 70 degrees centigrade suggests that there might be a common mechanism for the ability of water to resist a degree of thermal agitation.

Another theoretical problem is created by the results of transmission experiments, both direct and indirect: even assuming that some information about active molecules might be transmitted by a low frequency electronic amplifier, how can such information be detected in the midst of enormous electromagnetic noise? So far, this second puzzle has created a mental block for the few physicists who have been able to go beyond the accusation of fraud.

For information transfer to occur in transmission experiments, a mechanism must exist to enhance the signal-to-noise ratio. In the last decade, such a mechanism has in fact been observed, where the proper amount of noise can greatly enhance a specific periodic signal instead of overwhelming it; this phenomenon, based on the mathematical properties of chaos, is known as 'stochastic resonance'. I do not know whether the theory of stochastic resonance does indeed apply to transmission experiments. The only thing I am sure of is that, as long as scientists keep repeating 'impossible' instead of looking for an explanation, transmission experiments will indeed remain a mystery.

As suggested in Chapter 1, the potency of homoeopathic dilutions might be related to other anomalies involving the same ingredients of water, biological cells and long range interactions. The same might be true of transmission experiments. According to Benveniste, these experiments simply reproduce artificially what occurs naturally in cellular water: the transmission, amplification and storage of a molecular signal.

It will take more than scientific excommunication to do away with a signal that seems to be transmitted, stored and played back by water. For the moment, however, my impression is that the story of transmission experiments is in many ways a repetition of what happened after 1988 with high dilutions, in that an important piece of scientific research is getting bogged down by the burden of proof imposed by those who refuse to take it into account. The rest of the book is therefore devoted to an analysis of the behaviour of scientists when they are faced with evidence which they do not understand.

To conclude this first part of the book, I must emphasize that the essential point of my testimony so far has not been to convince anyone of the scientific interpretation of the phenomena reported by Benveniste, but simply to show that something potentially important has been found which deserves serious consideration. Perhaps the tentative interpretation suggested above will finally have to be modified, or even abandoned. Time will tell – provided that one gives the phenomena a chance.

PART TWO

The Strange Behaviour of Ordinary Scientists

INTRODUCTION

IF I had limited my testimony about the memory of water to the technical reports presented in the previous chapters, the whole story might well appear like a tale told by an idiot and signifying nothing. In fact, the story does make sense and neither Benveniste nor his opponents are crazy, even though they failed to communicate properly. It would be naïve or dishonest on my part to deny that, during my enquiry into this affair, I sometimes had a strong feeling of indignation when faced with the behaviour of my fellow scientists. What I have tried to do, however, is to go beyond these feelings in order to reach some understanding of the situation.

The purpose of the following chapters is to provide markers and intellectual tools to enable an interpretation of the sequence of events. In the so-called 'Benveniste affair', as in similar affairs, concentrating one's attention on specific people tends to blur the general (i.e. social) significance of events. Although the characters in the story are individuals, their names are of lesser significance because they have usually acted as representatives of some scientific institution. Even individual scientific articles are usually significant of the attitudes of the scientific community, because no article can be published without the *imprimatur* ('licence to publish') of some scientific authority. Since my purpose was to stress the general significance of the events that I was analysing, I have chosen to group these events thematically rather than chronologically and sequentially; I hope this choice will help the reader to interpret similar affairs that are bound to occur in the future.

In ancient Greece, when a runner brought some bad news such as the loss of a battle, he was sometimes put to death. Contemporary killings in reaction to a painful message are less cruel because they are of a more symbolic nature. The ancients, however, had one advantage over us: although they killed the messenger, at least they did not negate the meaning of his message.

When reading my report about the memory of water and about the way the evidence presented by Benveniste has been received, some people may feel that I have used too strong a language. I have used strong words indeed, but I think that they are not inappropriate to the situation. In order to appreciate this situation, one has to know that most of the evidence that was presented in Chapter 2 about high dilutions was published in 1988 but has not yet been discussed in any scientific review article. I feel that, in order to be heard above the clamour of those who have worked to denigrate this evidence, one must call a spade a spade. Hence my use of words such as 'censorship', 'mock attempts at duplication', 'perverse use of technical tools' and so forth in the chapters that follow.

Censorship implies a constraint, whose purpose is to prevent people from knowing something, allegedly 'for their own good'; such is the case of the memory of water, in which efforts were made to protect the populace from 'homoeopathic delusions'. However, I do not believe in a Manichaean dichotomy between those who supposedly possess the evil power to prevent others from gaining access to knowledge and those who are supposed to yearn for that knowledge. Rather, in the situations that I describe, I think that censorship, self-censorship and sincere support of conventional norms all play a part. This link between power and conformity was analysed by an eighteenth-century writer long before Marx defined his concept of alienation: 'Once you have thus formed the chain of ideas in the head of your contemporaries, you can then claim that you are driving them and that you are their masters. A stupid despot can force his will upon slaves with iron chains. A true politician binds them much more strongly with the chain of their own ideas.' As I will try to show, scientists are strongly bound by the chain of their own ideas, perhaps more so than ordinary citizens.

In Chapter 1, I indicated that, as a phenomenon occurring in the *physical* world, the memory of water would not shatter scientific knowledge, although it would probably lead to changes in our views about some aspects of molecular interactions within living matter. As we shall see, it is as a phenomenon occurring in the *human* world that the memory of water is perceived as threatening. Indeed, it threatens rigid frontiers between academic disciplines, the conventional pecking order between these disciplines and, last but not least, science as a sacred cow.

After the French version of my book came out, some colleagues told me: 'You talk about censorship, but you have not convinced me that

Benveniste is right. As long as his results are not proven, you should not use the word censorship.' Such reactions illustrate how hard it is for scientists to face the issue of censorship. Irrespective of the final status of the memory of water, the active intervention of scientists with the aim of stopping innovative research and of preventing serious discussion about that research is indeed a piece of scientific censorship. In other words, the issue analysed in the second part of the book is not the truth of any particular scientific finding but the freedom of research. This is far from being a mere academic issue to be settled among scientists. As I will try to illustrate in the next chapter, it concerns all of us.

CHAPTER FOUR

BE QUIET, THE EXPERTS ARE NOT WORRIED!

THE ROLE OF EXPERTS IN SOCIETY

We often show a healthy scepticism about expert knowledge on issues concerning the economy, politics or judicial matters. In scientific matters, on the other hand, we tend to be very credulous. I think that it is actually naïve to believe something just because it has been printed in a scientific journal such as *Science* or *Nature*. Both experts and lay persons are beginning to challenge the wisdom of putting too much trust in scientific expertise when that expertise has a direct impact on human affairs. In particular, a blind trust in scientific experts threatens democracy and public health.

Ecological issues furnish many examples of a challenge to scientific expertise. The so-called Heidelberg manifesto illustrates a conflict between those who claim a monopoly on knowledge and those who challenge that monopoly. In that manifesto, eminent scientists wrote: 'We fully agree about the goals of a scientific ecology centred on taking into account natural resources, controlling them and preserving them. However, through the present text, we strongly recommend that this taking into account, this control and this preservation be based on scientific criteria and not on irrational prejudices.' In fact, these scientists were manifesting a fear of losing their monopoly over rational knowledge – hence their accusation of irrationality against those who challenge this monopoly. As we shall see, scientists also have their share of 'irrational prejudices', specially when they claim to be fighting against obscurantism.

Others have taken an opposite stand and consider that the choice is not between a blind trust of notables and demagogy. They argue in favour of a different type of expertise, one that would be 'public, varied and open to contradictory debate'. This presupposes a fair amount of tolerance in the face of diversity of viewpoints.

In a recent article about risk management, one expert expressed his opinion in the following way:

> Lay people have different, broader definitions of risk, which in important respects can be more rational than the narrow ones used by experts. Furthermore, risk management is, fundamentally, a question of values. In a democratic society, there is no acceptable way to make these choices without involving the citizens who will be affected by them. [. . .] The alternative of entrusting policy to panels of experts working behind closed doors has proved a failure, both because the resulting policy may ignore important social considerations and because it may prove impossible to implement in the face of grass-root resistance.

What I find interesting about these quotes is not so much their content as their origin: they are taken from a long article that appeared in the *Scientific American*. The fact that these opinions could be printed in one of the leading journals of the scientific establishment indicates an interesting new trend in the delicate question of scientific expertise.

The issue of health hazards seems to me a most appropriate one for making a crack in the wall of scientific dogmatism, hence I shall use it to illustrate the risk associated with putting too much trust in official expertise. Except for the last one, the illustrations that follow have no direct connection with the memory of water. My purpose is simply to illustrate generally the danger of placing the burden of proof exclusively on the 'whistle blowers'.

The point is not to deny that campaigns about hazards sometimes turn out to be false alarms. One such example concerns predictions of ecological risks to forests, in which Swiss scientists have recently acknowledged that their past predictions about the death of forests have not been realised. However, it must be remembered that knowledge that worries had been overestimated is usually available only with hindsight. When human health is at stake, I think that one should be careful not to use scientific caution as an excuse for dogmatism or intellectual arrogance. In a controversy about health hazards, the crucial issue is not to minimize the risk of being proven wrong; what have to be estimated for each possible choice are the potential benefits of this choice (assuming that it is the correct one) balanced against its potential risks (assuming that is the wrong one).

The case of medical tests can serve to illustrate this point. When testing for an illness, two potential types of errors may occur: 'false negatives' and 'false positives'. Which of these is most dangerous depends on the specific circumstances. Generally speaking, however, it is more dangerous to let a serious illness go undetected than to make a false positive diagnosis because, in the latter case, the test will usually be repeated or some other test used to independently confirm the diagnosis.

In conventional views about health hazards, a distinction is made between risk *assessment* and risk *management*. Most experts concede that citizens need to be involved in risk management. However, they claim a total monopoly on risk assessment. This view is based on a rigid division between scientific knowledge and its social consequences that avoids crucial questions such as: 'Who chooses the experts?' and 'When experts differ (and this is often the case), who decides on a course of action?'

An important point to bear in mind is that of 'conflict of interest'. This concept is often used in circumstances where financial interests might bias decisions, but, as I hope to show in the rest of this book, this is too narrow a view of it. When a medical or scientific claim is made that contradicts conventional knowledge, experts are faced with other types of conflict of interest; whenever they express their views on unorthodox findings, experts are, by the nature of the case, both judges and judged.

THE TRAGIC STORY OF IGNAZIUS SEMMELWEIS

The concept of asepsis is usually associated with the name of Louis Pasteur; in France, there is hardly any town without a street commemorating his name. In fact, the first scientific demonstration that some fatal disease could be avoided by appropriate measures of hygiene was made by a Hungarian medical doctor called Ignazius Semmelweis. This fundamental discovery was published more than 20 years before Pasteur identified microbes as the agent of infectious diseases. However, the issue here is not one of scientific priority, but of scientific dogmatism and its potential dangers.

In the Semmelweis affair, as in many other affairs of this sort, the 'logic' of the establishment is very simple. It consists of the attitude: 'We don't understand, therefore it does not exist', or, more politely: 'We have not yet been convinced by your evidence.' For more than 20 years, the warnings of Semmelweis were not taken into account. Possibly a million deaths caused by puerperal fever during this period could have been prevented if this kind of scientific 'caution' had not paralysed the medical establishment. I cannot think of any better illustration of the potentially murderous effects of medical and scientific dogmatism.

Semmelweis was an obstetrician working in a large hospital. Contrary to other doctors, he gave his full attention to a well-known fact that his colleagues chose to ignore: women were more prone to die of childbirth in his hospital than those who gave birth at home. Motivated by his desire to change this situation, he finally found a solution, but failed to explain how it worked: he simply washed his hands with a substance that is now known to be an antiseptic. In retrospect, the fact that hospitals were more dangerous than homes can easily be understood: doctors killed the women through the germs which they carried from patients. After Semmelweis had discovered the way to reduce drastically the risk of mortality, he claimed that his clients were less likely to die of puerperal fever than those of his colleagues. Such a claim was not well received by these colleagues.

The story ended tragically, both for Semmelweis and for the women he was trying to help. He had demonstrated in a rigorous experimental manner that female mortality was correlated to a single factor: whether or not the doctor washed his hands. Although the remedy he proposed was simple (and certainly innocuous), he was ridiculed. Since he could not produce any explanation of his findings, nobody listened to him. Day after day, Semmelweis was the helpless witness of deaths which he knew could be avoided. He grew increasingly aggressive, to the point that he finally became insane. Although he had lost his sanity, he unconsciously gave a last illustration of the reality of the microbes which he had been fighting for so long: he died from an infection contracted through a small wound.

Semmelweis was eventually proved to be right, but his rehabilitation was discrete. The best way to honour Semmelweis would have been to relate his story in medical textbooks, as a warning against intellectual arrogance. What happened instead was that a statue was erected in his home town with a plate indicating that he was a benefactor of humanity.

IF IT WERE TRUE, SCIENTISTS WOULD SURELY KNOW ABOUT IT

The attitude characterized by the phrase 'I don't understand it, therefore it is impossible' can first of all be interpreted as an expression of mental rigidity, an idea on which I shall elaborate in later chapters. It can further be understood as an expression of faith in the scientific community. A sociologist of science named Westrum has illustrated this contemporary form of religious faith with an example that I find particularly interesting, in that it demonstrates the role of subjective factors in the evaluation of objective evidence. It also provides a spectacular example of a situation where the contribution of ordinary citizens did not concern risk *management* but risk *assessment*.

The battered child syndrome was first described in 1860. Three-quarters of a century later, a paediatrician specializing in radiology started a study based on X-ray photographs. Eight years after starting his study, he published a technical article describing six cases where photographs showed that the children had been beaten. Other similar cases were then published in the medical press but went practically unnoticed. It was not until 16 years after the publication of the first X-ray photographs of battered children (i.e. 24 years after the discovery of the first case associated with such photographs) that the American Academy of Pediatrics published the results of a national study concerning 749 cases of battered children. Thanks to the media, popular reactions were very strong and the number of cases mushroomed. By 1976, what had begun as a handful of cases was a massive phenomenon concerning hundreds of thousands of children in the United States.

In his description, Westrum emphasizes the impact on scientific knowledge of what is socially perceived as plausible or implausible. However, one other detail of the story of the battered child syndrome illustrates how far scientists can go to avoid facts which contradict their current beliefs: when the number of cases became too large to be ignored, a biological theory was first proposed to avoid facing reality; according to that theory, the cause of bone fracture was not beating but some biochemical deficiency leading to a special bone fragility.

When a doctor now teaching public health in a Paris medical school explained to me how he was required to adhere to this theory as a medical student, I was reminded of a similar case of scientific blindness. Some US scientists are so anxious to deny social causes of violence in their country that they have developed a biological theory of social violence. I recently read an article about the importance of genetic research on human violence written by the chief editor of *Science* (the official journal of the American Society of the Advancement of Science). How a genetic anomaly might some day be relevant to deal with murders that are 10 times more frequent in the US than in most other countries is a mystery to me.

Since I am interested in drawing parallels with the memory of water, I also want to make the following point: in order to see, one must look carefully. In the case of the memory of water, this obvious point seems to have escaped the members of the Scientific Counsel of INSERM when they first attempted to censor Benveniste's research on high dilutions. In a statement approved by all members but one, they claimed that Benveniste should have performed other experiments 'before asserting that certain phenomena have escaped two hundred years of chemical research'. In this respect, these INSERM notables were repeating the error of certain positivist scientists of the nineteenth century who naïvely believed that everything of any importance in science had already been discovered. In point of fact, Benveniste was not claiming to be the first to have noticed this phenomenon; far from claiming priority in his *Nature* article, he had quoted two references to previous work by others on high dilutions. It seems that, like most other opponents of research on high dilutions, the INSERM scientists had not even read the controversial *Nature* article on basophils that they were criticizing, as not only had the article quoted other teams who had detected effects of high dilutions on other biological systems, it also showed that Benveniste's team had already published work on high dilutions that had nothing to do with the staining of basophils.

In his analysis of scientific practice, Westrum has also given examples of the tendency of experts to believe that, within their own field, nothing of any significance could escape their attention and to assume, as soon as they reach a certain level of authority, that 'If it were true, I would surely know about it.' An instance of this type of scientific naïvety was apparent when experts who had been designated by INSERM came to check on Benveniste's transmission experiments. At the beginning of the visit, the expert on heart physiology expressed his doubts concerning the reality of the effects of high dilutions by declaring that *he* had never observed such effects. The institutional context prevented me from asking him some obvious questions such as: had he really looked and how? Instead I simply suggested a collaboration with Benveniste's team on the question of high dilutions. The expert said that his colleagues would probably not agree and, besides, he would need financial support from INSERM before embarking on such a project. Later on, Benveniste specifically requested a collaboration with him and he refused.

To complete this rapid overview of scientific short-sightedness, I should mention the following variant, which is often used as an alibi for a lack of scientific curiosity, namely: 'If it is true, it will finally come out.' The circular nature of such a statement can be demonstrated by analogy with judicial error. If one were to restrict judicial errors to the cases that have been acknowledged by courts, the above statement appears to be true but with such rare occurrence that it becomes devoid of any significant content. If, on the other hand, one considers all possible cases of errors, the statement appears naïvely optimistic. In France, for instance, sentences have been reversed only four times in the history of criminal trials. In the case of science, we find the same tautological confusion between truth and what scientists believe to be the truth. The trick to achieving this confusion is quite simple: one does not talk about truth but about 'scientific truth'. The history and sociology of science suggest that, when an environment is hostile to or unready for radical discoveries, the chances of these discoveries ever emerging in public are very slim. If by chance such a discovery should appear, its probability of survival would be even smaller. Hence the observed phenomenon of major scientific discoveries being preceded, sometimes several times and by many years, by the same discoveries made by pioneers whose names and findings vanished into oblivion.

Although the naïve view of scientific knowledge is still dominant, it is now beginning to be challenged both inside and outside the scientific community. An interesting example of a warning about fetishistic views of science was provided by Curien, a former physicist, who was at the time of the Benveniste affair the French Minister of Research and Technology. In an effort to cool passions about it, he stated in an interview to a major newspaper:

> If he tried to enter the CNRS (the French National Centre for Scientific Research), God would be flunked. He did an interesting experiment but no one ever succeeded in duplicating it. He explained his work in a fat publication but that was a long time ago and it was not even in English. He has not published anything since.

The author of the above statement must have known what he was talking about; before becoming a Minister of Research, he had been the General Director of the CNRS.

UNORTHODOX CANCER RESEARCH: THE CASE OF ANTONIO PRIORE

When Yolène Thomas, the biologist working on transmission experiments with Benveniste, submitted a project including a section about them, she was told that this was just another Priore case. At the time, she had never heard of Antonio Priore. The Priore case is described in two books that illustrate what I have been trying to tell Benveniste about his attempts to convince other scientists of the reality of transmission effects: support by a few established scientists does not guarantee that unorthodox research will receive a fair hearing. Should Benveniste succeed in convincing the French Nobel physicist Georges Charpak of the validity of his experiments (*see Appendix 7a*), his troubles would still not be over.

One of the books about Priore was written by Bader, a biologist who held the position of Scientific Director of INSERM and who was also one of the experts designated to deal with the Priore affair. The other book was written by a medical journalist. Both agree on the essential aspects of their story, which are damning for the scientific establishment. The Priore story contains some of the same ingredients as my story of the Benveniste affair: a refusal to examine evidence for lack of 'acceptable' theoretical explanations, suggestions of fraud and a tendency to ignore independent positive evidence, with escalating demands for further 'proof'.

Antonio Priore was an Italian immigrant living in the French town of Bordeaux. As a radar technician during the Second World War, he accidentally discovered that an orange forgotten in a submarine near electronic equipment had remained intact instead of rotting. After the war, as a result of reading a book which mentioned the effect of electromagnetic fields on cancer, he oriented his research towards cancer. He started experimenting on animals and progressively developed a complicated machine which produced various types of electromagnetic fields. With the help of a few doctors, he even experimented on some human cancer patients who were considered hopeless cases. The rumours of his successes attracted the attention of various medical people of his town.

Dr Berlureau, a veterinarian who was at the head of the Bordeaux slaughterhouse, also had a private practice. When people asked him for mercy shots for their pets who had cancer, he took them to Priore instead. In 1953 he obtained direct experimental proof of the efficiency of Priore's machine by sending tissues of the same animal before and after treatment with the machine to be analysed by a Paris cancer laboratory.

In 1960, two members of the teaching staff of the Bordeaux medical school, Dr J. Biraden and Dr G. Delmon, became interested in Priore's machine. They obtained some rats with cancerous tumours from the main cancer laboratory of the Paris area and had them subjected to treatment. The results were impressive: Priore succeeded in curing experimental T_8 tumours, that is, tumours of a type that nobody had ever before succeeded in curing. However, the results were not published until 6 years later, a delay that seems to have been due to opposition from Dr Biraden's boss. Dr Biraden was told that, if he published his research with Priore, he would not become associate professor; he therefore waited till his post was secure to publish his controversial findings.

Impressed by the results of Biraden and Delmon, Professor Guérin, the 'inventor' of the T_8 tumours, sent his assistant Rivière to Bordeaux to repeat the experiments. The allegedly invincible T_8 tumour responded to Priore's treatment. In 1964, a note was presented to the French Academy of Sciences by Courrier, who was not only a member of that Academy but also one of its permanent secretaries. The summary of this note reads as follows: '*Cancer research*. "Effect of electromagnetic fields on T_8 tumours grafted on rats." Rats with a T_8 tumour were submitted to electromagnetic fields at various stages of development of the graft. When the doses are sufficiently high, a complete regression of the tumour is observed as well as a total disappearance of the metastases that usually accompany it.'

After presenting this note, Courrier involved his personal assistant in Priore's research. He also asked an immunologist from Bordeaux, Professor Pautrizel, to help. Again, the results were positive. This was the fourth systematic experiment performed with Priore's machine and all had been positive. Courrier concluded his presentation of this fourth experiment in the following manner:

> Such results are surprising and can provoke scepticism. They were criticized even before they were known. A new claim is certainly always suspect. But, before condemning it, you have to check it first. This is what I have done, at the request of M. Rivière. I sent to Bordeaux 18 rats grafted with lymphosarcoma 347 on 25 January 1965.
>
> 10 control rats, 4 rats exposed 1 hour a day and 4 rats exposed 2 hours a day. It is difficult to use more than 8 rats at a time because you can place only 2 of them at the same time under the apparatus. The experiment represented 6 hours of exposure each day. It started on 30 January. My assistant was the only person to touch the animals during the experiment. These animals spent the night in locked cages located in the laboratory of Professor Pautrizel, at the medical school. Every morning all the rats were transported to Floirac. The 8 experimental rats were placed in the machine and were constantly watched by my assistant, Mrs Collonge.
>
> Results: Beginning of the experiment on 30 January. 15 days after the graft,

the last control rat dies. None survive. On 18 February, 19 days after the graft, the last of the 4 rats that were exposed 1 hour per day dies. None survive. On the contrary, the 4 rats that had been exposed 2 hours a day are in good health. They are back in my laboratory at the Collège de France. They are females and their vaginal cycles have been left intact.

There will be discussion about these results, which is a good thing. I consented to present these notes to the Academy for two reasons:

1 *When the issue is as serious as cancer and some light is being seen, it is our duty to seek what this light represents. We are not allowed to blow it off before knowing what it is worth.* [my italic]
2 [. . .] In the United States, biological effects of magnetic fields are actively being investigated. In specialized institutes, research is being done on the influence of such fields on tissue cultures, on microbes, on plants, on diastasis, on certain tumours. Results obtained so far with grafted tumours seem less important than those presented just now.

Some biological insight into the effects of the Priore machine was obtained by Pautrizel through a study of immune defence mechanisms unrelated to cancer. In his description of this aspect of the research, Bader writes:

As far as R. Pautrizel was concerned, he hoped to prove in a manner that could not be contested the authenticity of his first results. In a way, he succeeded since, in July 1970, Pr Aigrain (who was then the head of the National Office of Scientific and Technical Research) estimated that: the study conducted under the control of a bailiff has shown in an irrefutable way a stimulation of the defence mechanism of mice and rats against trypanosomiasis, perhaps by reinforcing these defence mechanisms.

This was also the opinion of Pr Lwoff who, after being sceptical for a long time, had come to Floirac and had been a witness to the interest of the research going on [with Priore's machine]. In fact, some time before, Pr Lwoff had asked Pr Avrameas, an immunologist of the Paris Pasteur Institute, to do some experiments with Priore's machine . . .

In the conclusion of his book, Professor Bader says that while writing his book he had been haunted by the words of Professor Jean Bernard: 'Don't contribute to doubts about science.' In the end, it was the restricted view of science which prevailed rather than the positive attitude towards exploration adopted by the Secretary of the Academy of Science and, although he had succeeded in convincing eminent scientists (including Lwoff, a future Nobel laureate) that his machine had spectacular biological effects, Priore died without any scientific recognition of his work. Since nobody else knew how to operate his machine, research on it stopped as well as the 'doubts about science' it engendered.

Lest the reader should think that scientific rigidity is a privilege of backward countries like France, I mention a case of unorthodox cancer research that was buried in the United States by scientific dogmatism: that of vitamin C and cancer. The story of this burial has been described by a sociologist of science in a book called *Vitamin C and Cancer: Medicine or Politics?* The medical issue was not as important as in the Priore affair. What was at stake was not a possible cure for cancer but an improvement in some specific types of cancer. However, the story is very significant because Dr Cameron, the initiator of this research, succeeded in impressing the Nobel laureate Linus Pauling. Pauling then associated himself wholeheartedly with Cameron in his fight against the cancer establishment. In spite of this prestigious support, the unorthodox research on vitamin C and cancer provoked the same type of reactions as does other unorthodox research, that is, mock attempts at duplication, hints of fraud and refusal to give this research a fair chance.

UNWANTED KNOWLEDGE: AN EXAMPLE RELATED TO THE MEMORY OF WATER

One afternoon of June 1992, I was with two other people watching a demonstration of Benveniste's transmission experiments. These experiments consisted in comparing the biological potency of two kinds of liquids: a control liquid (which in principle should be biologically inert) and an experimental liquid, which was identical with the control liquid except that it had been treated in the way described in Chapter 3 (*see page 38*).

In order to make his demonstration more convincing, Benveniste proposed that we should code the tubes he had just prepared in front of us. We went into an adjacent room and changed the labels identifying the tubes. After we had brought back the tubes, the operator started measuring the effect of their contents on the heart flow of two guinea pigs. While this was being done, Benveniste was watching the display of various other mechanical parameters on a screen. With confidence (too much as it turned out) he announced: 'This one is an experimental tube.' I was standing behind him and felt ill at ease, as I knew that the code of the tube indicated that it was in fact one of the dummies, which should have been inactive. This incident prompted me to propose a coding procedure designed to avoid this type of situation. With the new procedure, *nobody* would know the code before it was broken, not even the people who performed the encoding.

This type of incident (where a control tube turned out to be active) occurred several times. In a few cases, when the heart was unusually sensitive, it even happened that the contents of a control tube stopped the heart altogether. Through a systematic study of this suspected contamination problem, Benveniste found that active 'control' tubes contained liquid from specific origins. Two potential sources of apparently biologically active control liquids were identified: (1) phials of distilled water of a certain brand sold in pharmacies, and (2) physiological serum used by French hospitals and stored in glass bottles. (Conversely, French serum stored in plastic bottles and most foreign serums were found to be inactive.) Not all liquids coming from the two sources indicated above turned out to be active, but the difference with other (inactive) sources was apparently significant.

It was gradually deduced from various experiments that this unexpected activity observed could be caused by high dilutions of endotoxin. Endotoxin is a common bacterial product whose toxic effects on hearts are well known and can cause septic shock. It would be most surprising if liquids that had been specially prepared for medical use contained any measurable quantity of a well-known toxic product, and indeed they did not. However, it seems as if there was a detectable biological activity even in the absence of any detectable molecules. It is possible that this unexplained biological activity could be an example of the memory of water, as the behaviour of the 'contaminated' liquids mirrored that of the high dilutions of previously tested substances when they were heated for an hour or two at 70 °C; that is, their biological activity disappeared.

Although the exact nature of the contamination was uncertain, as well as its potential effects on the health of patients, Benveniste decided not to wait beyond November 1992 to warn the authorities that a potentially dangerous phenomenon needed to be explored. As a temporary safety measure he also proposed the heating of liquids from all suspect sources before any medical use. The analogy with the recent AIDS scandal of contaminated blood in Paris was of course immediately apparent to everyone and I was certain that, despite their tendency to misuse scientific 'caution', this time the authorities would act quickly. In fact, the sequence of events of the following 2 years showed that I had grossly underestimated the ability of scientists, doctors and politicians to ignore unwanted knowledge.

Starting in September 1992, the list of attempts made by Benveniste to warn various authorities of a potential health hazard runs to a few dozen items. For 2 years, these authorities ignored the warnings or used stalling tactics. This, of course, is also a political problem; however, I shall limit my examples to instances that are directly related to the theme of this book, that is, censorship in science.

The effects observed by Benveniste concerned hearts that had been sensitized by immunization. Extrapolating from animals to humans, it is reasonably certain that people who have not been inoculated in a similar manner to the guinea pigs would generally be in no danger. However, some categories of people face a possible danger of heart shock; in his official letter of warning (reproduced in Appendix 4), Benveniste mentioned vaccinated infants suffering from an infection, and in his letter to the editor of the major British medical research journal *The Lancet*, he mentioned other possible victims:

> Whatever the nature of ELA [endotoxin-like activity], its major *in vitro* cardiac effects must be dealt with, since, in man, the consequences of high doses of saline[1] are unclear. While harmless to subjects with a normal immune system, it could have adverse effects in those naturally sensitive to LPS [endotoxin], or mounting an immune response, or with immunodeficiency, cancer or haematological and/or infectious disorders.

The entire letter is reproduced in Appendix 4. The refusal of the editor of *The Lancet* to publish it contained no explanation whatsoever.

By way of contrast to the negative reactions related elsewhere, I shall quote one written positive reaction. In a letter to Benveniste, Professor Cabrol, the leading French heart surgeon, wrote the following:

> I have talked about it with Professor Gandjbakhch and indeed, after an operation, we do sometimes observe reactions which look like toxic shocks; but we have no explanation of these cases and the one you provide might be correct. We absolutely must go into it deeply, designate a committee of experts and clear things up.

Creating a committee of experts is exactly what the authorities did, but not in the positive spirit evoked here. Anthropologists have described ritual behaviour of so-called primitive people trying to exorcise some danger. In our societies, exorcism may go by the name of 'expertise'. The first step of this exorcism generally consists of the designation of experts who will work *behind closed doors*. More than 2 years after the first warning in the case above, the experts have yet to produce a report. Other facts also point to a desire to bury unwanted knowledge. Not only was Benveniste kept out of the committee of experts, but he was not even informed of its proceedings. For the past 3 or 4 years, he had been experimenting with the use of a complex and sensitive system to study unexplored biological reactions, yet these experts avoided all contact with the very person with the greatest practical experience! Whether positive or negative, the final report of these experts will be invalidated by the fact that they have ignored a major source of information on the topic of their report.

In Chapter 3, I mentioned the head of the National Institute for Health and Medical Research's violent reaction to the first report on transmission experiments. Even more surprising was the content of a letter that he wrote 8 months later concerning a visit of experts designated by INSERM. These experts were to evaluate a request for a research contract designed by Benveniste's team in an attempt to survive. In a letter sent shortly before the official visit to Benveniste's laboratory, the director of INSERM forbade him to ask one of the Italian physicists concerned with the theory of coherent domains to come and explain the aspects of their research that might help to establish the physical basis of the memory of water. He also gave strict orders to the expert visitors and to Benveniste to keep the question of contaminated serum out of the discussion of the proposed research contract.

From a strictly administrative point of view, the director of INSERM might have had the law on his side. The name of the Italian physicists did not appear in the list of scientists directly associated with the research proposal. It is also true that, in the section of that proposal devoted to the memory of water, the question of the contaminated serum had been left out. Nevertheless, if one considers the scientific and the medical stakes, this legalistic attitude seems inappropriate.

In a personal letter to the head of his institute, Benveniste mentioned an incident in which this scientist publicly said that he was a pain in the neck. I think that Benveniste does indeed belong in that category (as do I). The point, however, is that such a category of people can be subdivided into two groups. In the first one, you find people whose behaviour reveals only their personal problems, without any benefit for anyone; in the other, you find those who, through the expression of their personal problems, are socially useful.

To conclude this chapter on scientific expertise and health hazards, I would like to return to the first affair I examined in this chapter – that of Ignazius Semmelweis. In a biography of Semmelweis, a French author wrote:

> Imagine that today, in a similar way, some naïve person should come and cure cancer. He does not know what sort of a tune he would immediately be forced to dance to! It would be quite fantastic! He should be doubly careful. He better be warned of what to expect. He should be bloody careful to keep his nose clean!

This warning to scientific dissenters was written 70 years ago. As the next chapters will illustrate, it is not yet obsolete.

CHAPTER FIVE

DIRECT CENSORSHIP IS ONLY THE TIP OF THE ICEBERG

THIS chapter gives examples of direct censorship drawn from my study of the case of the memory of water. It must be emphasized that the direct use of power to suppress unwanted knowledge is only part of the process, even in a case as extreme as the one analysed in this book. To provide a context for the issue of censorship in science, I start by pointing out that this issue is itself suppressed by academic discourse, and then give examples drawn from other affairs, to show the multiplicity of ways used by scientists to say 'I don't want to hear about it.'

ACADEMIC DISCOURSE ON CENSORSHIP: A BLIND SPOT OVER A BLIND SPOT

The suppression of discussion about censorship might be as old as censorship itself. For instance, the *Encyclopedia Britannica* reports that: 'It is difficult to reconstruct the early stages of the controversy concerning the *Index* since arguments against the *Index* were themselves prohibited and destroyed, and the *Index Expurgatorius* was at first secret and not given general circulation.'

In the case of science, philosophers have oscillated between the tendency to consider science as the best model of rational thinking and the temptation to act as supreme judges of that rationality. More recently, sociologists of science have begun to analyse scientists' behaviour in a critical manner that casts strong doubts on scientific rationality. In their analysis, they stress the need to distinguish between what scientists think they are doing and what they actually do.

In the sociological analysis of science, the study of power relationships plays an important role. Nevertheless, the theme of censorship in science has remained an academic taboo. Using the *Social Science Citation Index*, I recently undertook a study of the way in which the topic of censorship appears in contemporary academic writings.

Academics know everything about postal censorship in imperial Russia or about censorship of pornographic films in India. On the other hand, the word 'censorship' is practically never associated with words like 'science' or 'scientific'.

The taboo concerning scientific censorship is most striking in articles that describe scientific censorship without naming it. The most spectacular example I found was an article entitled 'A crime worse than fraud threatens scientific progress'. Its author uses very strong language to describe scientific conformity to a dominant norm, referring to 'closed-minded stodginess prompted by suspicions of fraud', 'today's witch-hunting climate', 'the scholarly nit-picking critique, its emotion revealed in content devoid of feeling but clearly aimed at suppressing any new information' and 'the suppression of the quest for novel, invigorating Truth'. In spite of his violent criticism of scientific censorship, the author fails to name the concept he is illustrating.

Attempts to prevent discussions of censorship are less direct in the case of science than in the case of religious or of political censorship of the past. However, these attempts are still present, either in the form of a naïve belief in scientific 'objectivity' or in the form of coercion. In the case of this book, I have encountered both situations. Originally, I planned to write an academic text, whose outline I presented to a scientific publisher. This publisher was tempted but he hesitated and told me that he would consult an eminent scientist, whose 'objectivity he trusted'. I never heard subsequently from the publisher and decided instead to write a book for the general public. When the book was written, the director of INSERM, who had received galley proofs, tried to intimidate the publisher by invoking 'the rights of INSERM and his own concerning the consequences of the publication of this book'.

Serious work on the question of scientific censorship is rendered difficult by various obstacles, including the tendency of academic authors to shun controversial subjects. Testimonies about concrete cases of censorship are difficult both to obtain and to evaluate. People who see themselves as being hampered by some sort of censorship will tend to refrain from publicizing the fact. Generally, they will be afraid of being excluded from the category of serious scientists or of being considered paranoid. One major difficulty in the evaluation of a putative case of censorship is the following: how does one distinguish between the legitimate refusal to publish a scientific paper because it is 'bad' in the technical sense of the word and a refusal motivated by the belief that it is 'bad' in the sense that it goes against current orthodoxy? Since scientific censorship almost always hides behind technical reasons, this distinction is particularly difficult for a non-specialist to make. Another conceptual problem is that it is difficult to define violations of the 'laws' governing scientific behaviour, because these 'laws' are never explicit.

One might try to define censorship as an attempt to suppress 'scientific truth', but what exactly defines the 'truth' of a piece of research besides the fact that it has been published by a respected scientific journal and has then been quoted in a positive manner in other scientific journals? In spite of a vast scholarly literature on the human aspects of scientific research, most scientists lack any perspective on science as a human activity. Their relationship to the sacred literature (i.e. to the refereed journal) is akin to that of a fundamentalist who believes that *his* sacred texts give a true picture of the world. I emphasize this point because this tautological definition of scientific truth lies behind most implicit denials of the existence of scientific censorship.

One could attempt to bypass the problem of what constitutes scientific truth by saying that the criterion to be used is not the actual *level of truth* of any result but the *quality of the methods* used to obtain it. But this is only begging the question: who besides the gatekeepers of scientific orthodoxy decide on the quality of a piece of empirical or of theoretical research? As long as one stays within the either/or logic of true and false statements or that of valid and invalid methodology, the problem remains insoluble.

It is likely that blatant cases of scientific censorship are rare, however one chooses to define them. However, the important issue is not one of frequency. The fact that this frequency is apparently low simply reflects the efficiency of the scientific community in enforcing its norms without any obvious intervention. The important point is that current scientific practice (including the selection and training of future scientists) allows all types of research – except that most likely to lead to a genuine discovery.

In order to do away with scientific censorship, one would need to replace the ideal of truth with that of exploration. Lip service is paid to the need for scientific exploration by suggestions of allowing a small quota of controversial research. Others have proposed creating journals for scientific 'Miss Lonely Hearts'. As long as potentially novel research is considered at best as marginal or second-rate, scientific censorship will be part of the normal state of scientific affairs.

The central concern in my critique of scientific censorship is the protection not of a few scientists but of innovative research. Of course individual freedom is necessary, yet it is not sufficient. What is needed is not affirmative action to protect the rights of a few heretics, but a genuine re-evaluation of the nature of scientific innovation.

This type of re-evaluation cannot be achieved by a few radical scientists alone; it is also unrealistic to hope that the closed group known as the scientific community might spontaneously modify its norms. Such a re-evaluation implies some communication between those who stand inside the scientific citadel and those who are outside but whose life is nevertheless influenced by scientific dogmatism. In the last analysis, I think that the real taboo is not scientific censorship as such but a possible challenge to the rigid frontier between the inside and the outside of Academia. I think that, as long as knowledge travels in only one direction (from the inside to the outside), we will lose the battle against scientific dogmatism. The fact that this issue is still taboo is suggested by the manifestation of the following conditioned reflex: whenever the subject is broached, someone will try to stifle the debate by mentioning the spectres of Lyssenko or Mao.

CENSORSHIP AS PART OF THE NORMAL SCIENTIFIC PROCESS

In order to emphasize the fact that the various ways of rejecting unwanted knowledge are not specific to the memory of water, I shall illustrate some of them with examples drawn from other scientific controversies. I also wish to stress that editorial censorship is not the sole mechanism used to suppress unwanted knowledge, nor is it the most important one. Direct forms of censorship are only the visible parts of a system that, behind its apparent race for 'new' techniques, is fundamentally conservative.

Editorial Censorship

It is usually impossible to describe a case of scientific censorship without also describing the technical aspects of the case. I have been fortunate enough, however, to have first-hand knowledge of a case that can be described without having to do this. In 1980, I published a long technical article together with a medical doctor and a geneticist, in which was raised the issue of the validity of a certain type of consensus as a criterion of scientific truth. The case that we analysed in great detail was that of the links between genetics and schizophrenia. Our article was published by the major French journal of psychiatric research. Our next step was to attempt to publish this article in English. We sent the English version to *Acta Psychiatrica*, a scientific journal edited in Scandinavia. It was rejected, as is the fate of many scientific articles. What was unusual was the editor's adamant refusal to explain his decision. When I insisted, I was told that the criticisms of our article had been written by the referees in Danish, and would therefore be of no use to us. I responded: 'Never mind, I shall have them translated, please send them anyway', a request which was greeted by a flat refusal. This represents a case where a censor did not even attempt to hide behind technical arguments. The reason for the censorship can probably be found in the last paragraph of our 139-page article:

> For the moment, the very existence of such contradictions shows the potential danger of relying on expert opinion in matters of social importance. Disappointing though it may be, we feel that is wiser to admit that we know nothing, and that for the moment at least, genetics does not provide us with any real answers to the painful questions posed by the phenomenon of 'schizophrenia'.

Institutional Forms of Censorship

Like many other forms of repression, scientific repression is rarely explicit. Material evidence of this form of censorship is rare and published material is of course exceptional. I did not find any such material, but I did find a published text corresponding to an attempt to intimidate a research worker with unpopular views on science. In October 1990, *La Recherche* (the French equivalent of *Scientific American*) published a long article analysing the manner in which the low percentage of success of *in vitro* human fertilization was being artificially boosted. A few months later, a worker in the field who happened to be a former president of the CNRS wrote: 'I do not wish to start a polemical argument with the author since only the appropriate commission of the CNRS can pass a judgement on his writings. They will surely ask questions about the way a research worker can construct an argument.' If that was not an institutional threat, what was it?

Mock Attempts to Duplicate an Experiment

One of the most common ways of rejecting a possible discovery consists of attempting to duplicate an experiment in such a way as to render the chance of its success minimal. Such behaviour is usually not deliberate. Its victims often consider it to be evidence of bad faith or technical incompetence. My personal view is that it results from poor communication and a mental block on the part of those who wish to reject uncomfortable novel findings. Because a multitude of factors can contribute to the failure to duplicate an observation and because these factors are largely unknown when a new field is being opened, what I call 'mock attempts to duplicate' are among the most efficient ways of rejecting unwanted knowledge.

An attempt to duplicate an experiment has little validity if the authors do not start by becoming familiar with the process that they are examining and first try to follow the original protocol as closely as technically and humanly feasible. It should be emphasized that, from a strictly logical point of view, a negative result is more difficult to interpret than a positive one. An example drawn from everyday life can be used to illustrate this. Suppose that someone were to send you a recipe for preparing a cake. If you succeeded in producing a good cake, you could assume two things: (1) the recipe was a good one; (2) you followed it correctly. If, on the other hand, the cake was a failure, it might not be obvious whether you had missed out something crucial or whether it was a bad recipe. Now imagine you sent this recipe to five friends and that three of them failed while two succeeded. You might reasonably conclude from this that the recipe was indeed good, but difficult to follow.

Scientists are aware that most new experimental procedures are difficult to duplicate exactly. But their conscious awareness of this is often confined to those procedures with which they are familiar; when they examine someone else's procedure on the other hand, the awareness of the inherent technical difficulties may be very limited indeed.

Difficulties in interpretation can occur even in the 'hardest' and most clear cut of all experimental sciences, namely physics. The history of science shows that even apparently simple physical experiments often could not be duplicated, and that the consequent acceptance or rejection by other scientists of an apparent failure to duplicate was made not on the basis of the intrinsic quality of the attempt but according to prior expectations of those involved.

One of the major laws of electricity was established during the nineteenth century by an experiment presented by the French physicist Charles Augustin de Coulomb. As we learned in school, the force between two electric charges varies with the inverse of the square of the distance between them (the 'inverse square law'). However, recent attempts to duplicate the original experiment using a Coulomb scale have failed to produce results consistent with those expected from this law. In this case, the failures were interpreted not as failure of the law itself, but as due to the multiplicity of factors influencing an experiment, including the necessity of having a certain know-how and problems caused by the electrostatic properties of materials used in the manufacture of contemporary clothes.

An inverse example of the ambiguous status of duplication is provided by an experiment first performed by Isaac Newton. This experiment is apparently quite simple since it uses only a beam of natural sunlight and two glass prisms; when a beam of white sunlight enters a glass prism, it is bent and split into its coloured components. When a second prism is added, the angle of bending appears clearly to be different for each colour. For almost half a century, Newton's opponents claimed that this experiment could not be duplicated. In retrospect, however, such claims can be explained by the failure to use glass of the proper optical quality or to place the prisms in the proper alignment.

Biological experiments are usually more complex than those in physics and intrinsically difficult to reproduce. An interesting example of this (analysed by Behar) is the historical controversy over the transmission of nerve impulses across the synaptic gaps between nerve endings. It was found experimentally by Loewi that the carrier of nerve impulses across these gaps is not electrical but chemical. However, to make a long story short, orthodox scientists fought tooth and nail against this finding and 'demonstrated' that Loewi's experiments could not be duplicated. In fact, these detractors had used different animals from those used by Loewi. Finally it was understood that artefacts connected with the use of other animals and unknown of at the time had masked the chemical phenomena observed by Loewi. After two decades of bitter struggle, the chemical nature of neurotransmission eventually became the new official dogma.

Behar points out the analogy with Benveniste's transmission experiments. In the case of nerve impulses, messages that were thought to be purely electrical turned out to have a chemical component. If Benveniste and Thomas are right, the opposite would now be true: biochemical information that is at present considered to be located purely within molecules could in fact be transmitted by purely electromagnetic means, with no concurrent molecular transfer. As we shall see in the next chapter, Benveniste's detractors, like those of Loewi, have also made errors in their 'failure to duplicate' his high dilution experiments.

Scientific Harassment

I define scientific harassment as an avalanche of technical criticism whose goal is not the achievement of scientific knowledge but scientific fighting. The bellicose nature of many academic arguments is well known, both inside and outside Academia. In his book, *Science in Action*, the French sociologist of science Bruno Latour uses a large number of military metaphors to describe scientific argumentation. In a chapter devoted to scientific literature, he uses metaphors such as the following: mobilization, attacks, destruction, allies, dismantling, tactics, strategy, the weakening of enemies, helping allies who have been attacked, enemies fighting between themselves, positions of strength, opponents who are pushed back through the recruiting of an increasing number of allies, battlefields, propping up a fortress, supporters, successive defence lines, comparing a musket to a machine gun, fighting tanks with swords, courageous fights, isolated opponent, changing camp, alliances that are broken, head counts that are as inadequate in science as they are in an army, alignment of troops, competitors entering the lists, victory, torture, troops aligned and trained, choosing the terrain, besieging readers, the battle-front of the controversy, scientific texts written for defence or for attack, stronghold. The last item of this military list is the word 'bunker'.

Latour also analyses in detail the manner in which scientific proofs are constructed and the way that, instead of serving as a tool in the search for scientific knowledge, this construction is used as ammunition against competitors. Far from a simple literary exercise, Latour's descriptions appeared to me to provide an important key to what I was observing in the fight over high dilutions.

Finally I would like to make two points. First, although this bellicose spirit is common in scientific argumentation as a whole, the partiality of the major scientific publications is most obvious in cases that appear to threaten the establishment. The second point concerns the relationship between masculine stereotypes and scientific controversies. I believe that there is a link between what Susan Bordo calls 'the cartesian masculinization of thought' and the tendency of scientists to prefer power to knowledge. In plainer words, a scientific controversy might be an end in itself, because males enjoy a good fight. In a sense, scientific controversies are the modern equivalent of mediaeval tournaments; the main difference nowadays is the loss of the traditional male value of fair play.

Theoretical Objections

Although contemporary scientists do not say, as did a French positivist scientist of the last century, that 'Everything of any importance that could be discovered has already been found', many act as though it has. This unformulated principle guiding their behaviour has been described by a sociologist of science as 'ethnocentrism of the present'. Just as most people have difficulty in realizing the relative nature of their cultural values, contemporary scientists often forget that today's scientific truth may become tomorrow's past error.

The history of the optical laser is a recent example of such short-sightedness. Obsessed by the question of thermal equilibrium, scientists refused to consider it seriously as they decided that it would represent a theoretical impossibility according to the laws of thermodynamics. Some were sarcastic about it as they thought it corresponded to the absurd idea of a 'negative absolute temperature'. The last episode of this long saga was told by Hecht in a book entitled *The Laser Pioneers*:

> By the time he [Theodore Maiman] succeeded in making the ruby laser work for the first time, on May 16, 1960, he was not supposed to be working on the program. [. . .] His success is undisputed, but he almost immediately ran into problems reporting it. Hughes management reacted enthusiastically *once the laser worked*, and sponsored a full-fledged press announcement in early July. [. . .] A more serious problem came when Maiman submitted his paper for publication. The then new *Physical Review Letters* summarily rejected it as 'just another laser paper'. [my italic]

Physical Review Letters has since become the top international journal for physics. In 1988, it published the seminal paper of the Italian authors of the theory of coherent domains entitled 'Water as an electric dipole laser' (*see Chapter 1*), which remains unwanted knowledge for most scientists. What I find fascinating in the laser case is that the theoretical objections about the absurdity of 'negative temperatures' did not even correspond to the state of knowledge of the time. These arguments were based on classical thermodynamics, but these had been contradicted several decades before with the discovery of the fact that atoms obey different laws. Some of the implicit objections to the theory of coherent domains seem to be equally outdated.

Suspicions of Fraud

Suspicion of fraud is the ultimate weapon used by those who wage a war against observations that do not fit their world view. From a purely intellectual point of view, this represents an expression of weakness; from a practical point of view, however, it is an expression of power, and actually of an abuse of power. Like other tactics to dispose of unwanted knowledge, intimation of fraud is generally made in good faith and is based on some anomalous event, real or imaginary. The Priore case outlined in the previous chapter contains two such examples.

First, during a discussion between Parisian scientists, the biologist André Lwoff said that the electric meter seen in Priore's house was grossly insufficient to operate the machine Priore claimed to have developed. This was indeed true, but in fact electricity was also being delivered without going through his house meter. After discovering this, Lwoff changed his mind and became interested in Priore's research.

The second example illustrates the circular reasoning that underlies most suspicions, that is, the observations are impossible, therefore their report is based on fraud. In one experiment, a British team reported positive results obtained with inbred mice. Someone from the same laboratory, who had not participated in the experiment, started rumours of fraud because on their return from France the scientists discovered that they had forgotten to label a few of the mice and, in an attempt to identify them, had grafted tissues from the same inbred line. These grafts were rejected; however, instead of considering the possibility that Priore's treatment might have modified the immunological properties of the mice, some people concluded that this was proof of fraud. In fact, other positive results obtained under very stringent conditions should have been sufficient to invalidate this interpretation. Later work revealed that isografts were indeed rejected after animals had been irradiated with Priore's machine. In other words, what had been taken as evidence of fraud was actually a new and interesting fact in itself. While this was later recognized, the harm had already been done.

In the ordinary world, when someone makes an accusation, the burden of proof is on him, not on the accused. However, within the scientific community, scientists do not follow the same laws. When a French professor of law directed some graduate work on the Benveniste affair resulting in a report entitled: 'The history of a scientific trial; the case of the memory of water', she was horrified by what she discovered about the mores of scientists. On one occasion, Benveniste went to court in an attempt to stop some of the slander being used against him. By doing so he violated the unspoken golden rule of any closed group: never let outsiders intervene in our affairs.

The Misuse of Language

The fact that the stakes in the Benveniste affair go beyond Benveniste himself and beyond the memory of water or homoeopathic medicine is illustrated by the misuse of language. To illustrate this process, I have chosen another example that is even more extreme. This example is drawn from a book entitled *A Threat to Science* published by Odile Jacob. This publishing house is part of the French scientific establishment and is the publishing house preferred by well-known scientists writing for the general public. In a section entitled 'A Hitlerian conception of science', Evry Schatzman, who is a member of the French Academy of Science, wrote:

> The statement that science is the product of a dominant group is full of dangers. One can find in Hitler's phrases quoted by Rausching a radical form, perhaps caricatured, of the principle of consensus expressed in a different way by Kuhn, Feyerabend and Latour. In other words, when one goes from philosophic reflexion and discussions about the nature of science to a sort of vulgate, a popularization negating science, one finds again *quite simply* Hitlerian theses.
>
> Under the guise of anarchy (Feyerabend) or of leftism (the texts of *Pandore* edited by Bruno Latour) are being created the theoretical bases of a policy that has a painful kinship to Hitlerian policy.

This lumping together of various forms of historical, philosophical and social critique of science with Hitler's ideas about science reveals that Schatzman is completely incapable of facing up to any intellectual challenge to his idealized vision of science. In this particular case, the author himself gave a possible clue to this idealization: the observation that, as a young man trying to escape the Nazis' persecutions of Jews, he was saved by the fact that he was a brilliant science student, and the director of a French observatory in the South of France agreed to hide him. According to Schatzman, the disparate views of Kuhn, Feyerabend and Latour all have a 'painful kinship to Hitlerian policy'. I wonder who will prove sufficiently evil to be compared to me, a traitor to science and a Jew to boot!

My classification of various forms of censorship in categories and sub-categories is somewhat artificial, as there is hardly any form of direct censorship that does not also contain some shreds of technical arguments and vice versa. It may nevertheless be useful, because it emphasizes the diversity of means used to attain the single end of rejecting unwanted knowledge.

THE CASE OF THE MEMORY OF WATER: EXAMPLES OF DIRECT CENSORSHIP

Editorial Censorship

My first example, from the Benveniste affair, illustrates how shreds of technical arguments can be used to reject unwanted knowledge. In autumn 1988, 4 months after the publication of his report entitled 'High dilution experiments a delusion', the editor of *Nature* wrote the following statement in an article entitled 'Waves caused by extreme dilutions':

> My own guess is that Dr Benveniste's colleagues will now be counting basophils in replicate, following the standard procedure for controlling sampling errors, and will be eliminating unavoidable observer bias by making blind measurements a routine. I expect that the results will not differ from those obtained in the three blind experiments (each with two observers) at Clamart on 9 and 10 July;[1] it will be extremely interesting if it should be otherwise, but no doubt Dr Benveniste would prefer to publish that intelligence in some other journal.

As a matter of fact, Benveniste did duplicate his experiments, using only blind measurements. After the scandal created in France by the publication of *Nature*'s report, scientists of another INSERM laboratory observed and supervised two series of duplications of high dilution experiments. Until the end of the series, Benveniste and his co-workers did not know what they were measuring, since the code was devised and kept by scientists of this other laboratory. Contrary to Maddox's anticipation, these new series of experiments confirmed previous observations of the biological effects of homoeopathic dilutions. Also contrary to Maddox's prediction, Benveniste did try to publish his new results in *Nature*. However, in spite of his previous implicit offer to publish the results of high dilution experiments provided that they were performed blind, the editor of *Nature* refused to publish these new results. His letter to Benveniste contained only a single paragraph:

> Thank you for your fax.[2] It does not matter whether you withdraw your paper or we reject it – I'm afraid it is the second course that we would in any case have followed. The reasons are explained in the enclosed report of one referee. Briefly, as you will see, there appear still to be errors of a statistical character in your work.

Since Professor Spira, who had supervised the new experiments and co-signed the article, was a professional statistician, the accusation of statistical errors was somewhat surprising. I will illustrate the nature of these 'errors' by referring to the first two mentioned by the referee. The very first complaint was that he did not know whether numbers appearing in a table indicated basophil counts or percentages. In fact, the title of the table indicated that it was the former. The second 'error' was even more incredible. The referee complained that a statistical error associated with a basophil count was too large, in fact larger than the number itself. Actually, he had confused the error squared (technically called the variance) with the error itself! (The fact that the large number referred to was indeed a variance was explicitly shown in the table heading.) An undergraduate making such a gross error in a statistics course would most certainly be failed. Spira and Benveniste sent a rebuttal to *Nature*, but to no avail. *Science*, the major competitor to *Nature*, used similar tactics to refuse publication of these new results. In the end, they appeared in the journal of the French Academy of Science.

I have already mentioned that the British medical journal *The Lancet* refused to publish the short paper on the contamination of physiological serums (reproduced in Appendix 4) without providing any explanation. In the same year, Benveniste also sent a short paper on his transmission experiments as a contribution to a meeting organized by the American Society for the Advancement of Science, on the topic of 'Scientific innovation'. The organizers dismissed his contribution using a standard impersonal letter of the sort used to reject the work of a graduate student.

Benveniste protested and sent 10 affidavits from other scientists about the *bona fide* nature of the transmission experiments. He was then allowed to present a poster, as if he were a doctoral student presenting his thesis work.

When he attempted to publish his work on transmission experiments in a scientific journal, he was even less successful. In 1994, he submitted a detailed description of his transmission experiments to the *Proceedings of the US National Academy of Science*. In order to avoid the risk of having his article simply ignored, he had previously written to the president of this Academy, asking him for advice. The answer was a flat refusal contained in a one-paragraph letter: 'I have received your letter of January 19, 1994. Unfortunately, our Academy does not have a mechanism to deal with the type of issue you raise.' Journals of the establishment are indeed not prepared to deal with potentially important discoveries in an open-minded way.

Institutional Censorship

The first level of institutional censorship is intimidation, whether explicit or implicit. This type of strategy is most efficient when used to isolate the heretic from potential co-workers by threatening the positions of those involved. For instance, a young research worker who was trained both in physics and in medicine had written to Benveniste to inquire about his research on the memory of water. In response, I invited him to the seminar that I was organizing with Benveniste in spring 1993. Although he was deeply interested, the young man finally declined our offer. He was quite frank with me about his reasons, explaining that, as long as he did not have a permanent appointment within INSERM (the French Institute of Medical Research), he could not afford to associate with Benveniste.

Another case was that of a brilliant biologist who was classified as first of his group in the competitive examination to obtain a permanent appointment within INSERM. Unfortunately, a member of the jury noticed that he was planning to return to Benveniste's laboratory where he had once worked as a doctoral student. Although the young man's research project had nothing to do with the memory of water, the member of the jury who had detected a potential threat to official science said that he would resign from the jury if the young biologist was allowed to work with Benveniste. The young scientist was then reclassified as seventh in his category; since there were only six posts available in this category, he was not appointed that year. The following year, he tried again to obtain an appointment within INSERM, but not in Benveniste's laboratory. This time, his scientific merits were recognized and he obtained a permanent research position.

In the case of an established scientist like Benveniste, with a secure, tenured position within INSERM, the situation was different. Here his research was progressively smothered by gradually cutting down his budget and finally closing his laboratory. Without going into the details, I will illustrate this by mentioning four episodes.

In 1989, INSERM used the routine 4-year evaluation of its laboratories as an opportunity to try to censor Benveniste's research on the memory of water. In this instance, he was saved by the personal intervention of the Minister of Research (the man who thought that God would be flunked by the CNRS) and by that of the general director of INSERM, who did not follow his scientific council's recommendation. The head of INSERM maintained Benveniste as director of his laboratory on condition that he stop communicating with the general public for a 6-month probationary period.

The second episode (mentioned in Chapter 3) was that the director of INSERM attempted to prevent Benveniste from communicating information about his experiments, not only to the general public, but also to his colleagues. On this occasion, the general director threatened to 'draw serious consequences' from what he considered to be 'the pernicious character of such "information"'.

The third episode was the crucial one. From an administrative point of view, all INSERM laboratories have a maximum lifetime of 12 years, after which if they want to continue as a team, the scientists involved must present a new project in competition with others also asking for a laboratory. Because of the intimidation of INSERM scientists mentioned above, the few scientists who had agreed to join forces with Benveniste all belonged to another research institution, namely the CNRS (to which I also belong). For this reason, Benveniste was not even allowed to *ask* for a new laboratory. Swallowing his pride, he decided to request instead a short term contract called a 'Young researchers' project'; this type of contract is meant for scientists who have yet to show their ability to produce successful research. I have already mentioned the fact that the director of INSERM used formal administrative devices to prevent an evaluation of two aspects of the research on the memory of water, that is, the theoretical aspects and the medical aspects. In spite of these restrictions, the moderate conclusions of the commission sent by INSERM could have permitted an acceptance of Benveniste's request. Despite this, however, INSERM's scientific council took a very negative position, with one member going so far as to qualify Benveniste's research as 'black magic'. His request was finally turned down.

The fourth and final episode occurred in 1994, when Benveniste tried once more to obtain a limited research contract from INSERM. This time, all references to the memory of water were removed from the research proposal, and it was built entirely around conventional lines of research. This type of self-censorship did not prove sufficient. In the absence of last-minute changes, the institutional verdict against research on the memory of water is now final.

THE LINK BETWEEN CENSORSHIP IN SCIENCE AND THE 'GOLDEN RULE'

As with other human phenomena, censorship is the result of several 'causes'. The case of Galileo's condemnation by the Holy Inquisition may serve to illustrate this point. Here three types of 'causes' can be detected. First, on the psychological level, Galileo probably precipitated his condemnation by his arrogance and provocation of various authorities. Secondly, his cosmological views were the apparent 'cause' of his condemnation. However, in my opinion, the third and most crucial issue was that of authority. In other words, the challenge created by Galileo was: 'Who is the boss?' This challenge was clearly worded by him; as the following extract shows:

> Let us admit that theology is devoted to the highest divine contemplation so that, because of this dignified position, it occupies the throne among sciences. [. . .] its masters should not take it upon themselves to impose their decisions in controversies about disciplines that they have not studied and in which they have no experience. Indeed, it is as though an absolute monarch who was neither a doctor nor an architect but who had the liberty to command others, decided to administer drugs or to erect buildings according to his own caprice. His patients would be in jeopardy and his buildings would soon crumble down. . . .[3]

In Benveniste's case, the same multiplicity of causes can also be found, with psychological, intellectual and social causes all contributing to the suppression of unwanted knowledge. From a psychological point of view, his lack of diplomacy made things easier for his enemies. From an intellectual point of view, the scientific issues raised by the memory of water were apparently at the heart of the fight between himself and the establishment. Finally, as in Galileo's case, the issue of authority concerned personal power, but the stakes were also less personal. The implicit question posed by this case is: 'What kind of group is entitled to reach a final decision on certain controversial matters?'

In this case, I believe that the most crucial issue is the rigid separation between the inside and the outside of Academia. This separation is usually labelled 'academic freedom', meaning that: 'Everything which furthers the progress of science is intrinsically "good", while any attempt to put checks on that progress from the outside is intrinsically "evil".' I found a striking example of this opinion in an editorial published by *Nature* under the title 'Criminalizing research'. The subtitle of this editorial was 'West Germany seems bent on a restrictive law on embryo research'. What caught my attention was the fact that this denunciation of censorship *of* science immediately followed another editorial which initiated a case of censorship *in* science. Titled 'When to believe the unbelievable', that editorial concerned the article of Davenas *et al.* on high dilutions published in the same issue of *Nature*. As we shall see, it was the signal of a scientific witch hunt.

Another aspect of this separation is that: 'As long as it has not been validated by some inside authority, any information from outside the academic citadel is to be ignored.' In other words, scientists claim a sort of extraterritorial status that is reminiscent of the status of clergymen before the separation of State and Church.

The head of INSERM's behaviour illustrates the priority he gave to the above consideration. In 1989, his sole condition for maintaining Benveniste as director of his laboratory was that he should refrain from communicating with the general public. In 1994, in a letter to Benveniste about the closing of his laboratory, he wrote: '*Publish*,[4] and there is no reason why you should not be accepted once again.' In further comments on his version of 'publish or perish', the general director mentioned that this publication should be 'in a high level scientific journal'. In other words, scientists are not allowed to pursue any kind of novel research unless they have the prior permission of established scientists. The *imprimatur* of referees is the modern version of the *nihil obstat* of the Catholic Church.

The fact that the director of INSERM refused to take any personal responsibility for scientific censorship but hid behind requests for high level publications illustrates the close link between various types of censorship. Institutional censorship rests on editorial censorship, which itself rests on the types of censorship that will be illustrated in Chapters 6 and 7.

CHAPTER SIX

THE PERVERSE USE OF LEGITIMATE TECHNICAL TOOLS

THE forms of censorship illustrated in the previous chapter are not those normally employed by the scientific establishment in order to control its members. These direct forms are rather rare, especially that of institutional repression. I have begun with these extreme forms because they are easier to explain than are those forms of censorship that hide behind technical arguments. At first glance, these technical arguments appear to be legitimate tools of intellectual criticism and, in everyday science, technical arguments are, indeed, generally used in this manner. Only when normal science is 'threatened' are legitimate tools used incorrectly and unfairly. In such cases, the normal sequence of arguments and counter arguments is grossly biased in one direction.

When the scientist being criticized is considered to be heretic, the balance of power is tipped against him and technical criticism becomes a veil for censorship so that, using all his strength, the heretic can barely present his case. As we shall see, those who attack him may use any means, however irrational. Because these means are usually technical, this chapter is by nature also more technical. However, I have done my best to bypass the conventional rule which decrees that scientific arguments can be appreciated only by scientists.

MOCK ATTEMPTS TO DUPLICATE AN EXPERIMENT

Duplication can be used either to search for scientific knowledge or to run away from it. In other words, it is not the technical tool itself that I am criticizing, but the manner in which it is sometimes used. The case of high dilution experiments provides a typical example of the perverse use of legitimate technical tools. Within weeks of the publication of experiments about high dilution effects on the staining of human basophils, *Nature* published four reports claiming that these effects could not be duplicated. As we shall see, the authors of these reports operated in such a way that their *a priori* chance of success was minimal. In each case, they acted as if they desperately wanted to fail in their attempt to reproduce the results reported by Benveniste and his colleagues.

The high dilution experiments published by the INSERM group (together with scientists from three other countries) used a single biological test: the changes observed in the staining properties of human basophils. The three teams reporting their failure to reproduce the results of the INSERM group in fact failed to reproduce the protocol published by Davenas *et al.* In each case, the difference was not merely in minor details that may or may not be of crucial importance. The authors actually used a different test. Instead of studying the staining of basophils, they studied another biological reaction (for instance the emission of histamine).

One typical detail suggests that the authors reporting 'failures to reproduce' were actually none too keen on succeeding. In two of the experiments, the authors did not even use basophils from humans but from rats! The differences existing between rats and humans did not matter to them since the real purpose was not to study the possible existence of high dilution effects but to detract from its very possibility. The significance of this detail is compounded by the fact that one of the authors was Henry Metzger, a biologist who, on two occasions, had played a crucial role in the censoring of research on high dilutions. He was one of the referees who had strongly opposed the publication of the article of Davenas *et al.* and was also one of the two experts chosen by INSERM in 1989 to evaluate Benveniste's research on the memory of water. His negative report contributed to the decision of the scientific council of INSERM to censor Benveniste's research on high dilutions.

The last instant 'proof' of the fact that high dilution experiments cannot be reproduced was provided by the editor of *Nature* himself. When he finally published the heretical article on high dilution effects, he wrote an editorial inviting readers to detect flaws in Benveniste's report. Four days later, he came to Clamart to participate in the 'check' he had initiated. It was unheard of for the editor of a scientific journal to act personally as an expert on a series of experiments he had just published. In doing so, Maddox put himself more directly in the situation of being both judge and judged. The non-neutrality of his stance was revealed by the fact that, instead of bringing biologists to oversee the experiments, this former physicist came with two experts on scientific fraud.

Maddox spent 5 days in Benveniste's laboratory with Randi, a professional magician who had previously been called in to investigate cases believed by some to be explicable only by fraud, and Stewart, a physicist also specializing in the detection of scientific fraud. The role played by the suspicion of fraud in this case will be examined in the next chapter. For the moment, let us simply examine the report by Maddox of his failure to observe high dilution effects.

At first, the 'fraud squad' simply watched Davenas as she performed three experiments. The results of these experiments were positive. The observers then coded the tubes prepared by Davenas, so that the control tubes and the various high dilution tubes were randomly mixed. Again, the outcome of the experiment was positive.

Rather than taking the risk of being a witness to other such blind experiments (thus validating previous reports on high dilution effects), Maddox then intervened directly in the experimental protocol. He was apparently convinced that the positive results he had just observed could be due only to fraud. In the case of the fourth experiment, fraud could not have been perpetrated during the basophil *counts*, since they had been counted in a blind fashion. The implicit hypothesis of the report published by Maddox was therefore that fraud must have occurred during the *preparation* of the dilutions. In other words, it proposed that Elisabeth Davenas must have acted as an expert magician, adding aIgE in high dilution tubes under the nose of someone specialized in the detection of scientific fraud and while being watched by a professional magician, also an expert in this type of detection. Quite a feat indeed![1]

In order to obtain the 'proof' they were searching for, Maddox and his co-workers tightened their coding procedures. To make things appear more dramatic, Randi stuck the envelope containing the code on the ceiling. Stewart insisted on handling the pipettes himself during the preparation of the microscope plates, in spite of the fact that he had no experience in dealing with cells (he was a physicist).

The result of this new procedure was that one experiment was disqualified because the corresponding basophils were unresponsive even to high concentrations of aIgE, and two experiments gave negative results. Having thus obtained the failure he had been looking for, Maddox ended his investigation. He then published a highly critical report containing both implicit and naïve arguments. For instance, he wrote that he was 'surprised to learn that experiments do not always work'. Even a biologist who has never read a word about the sociological components of scientific duplication knows that biological phenomena are difficult both to control and to reproduce. This is precisely why the INSERM scientists had performed 300 experiments (including some 50 blind experiments) *before* being challenged by a fraud squad. There is also considerable biological naïvety contained in the expressed surprise of Maddox at the fact that the blood of some donors was unresponsive to aIgE, even with high doses;[2] if human organisms were not endowed with a high degree of individual variability, the human species would have been destroyed by epidemics a long time ago and there would be nobody left to argue about biological variability.

Maddox based his conclusion about the fact that high dilution experiments cannot be reproduced on two negative results which, in his opinion, superseded the large number of previous positive results. Two further points illustrate the extent to which Maddox was blinded by his desire to prove the non-existence of high dilution effects.

The first point concerns the pressure put upon the experimenters. During the preceding 130 weeks, Davenas had never performed so many operations in so short a time as she did in the week of the visit; in that week, she prepared seven series of dilutions, seven series of blood cells and performed an unprecedented number of basophil counts. These were hardly favourable conditions under which to perform difficult experiments, even without the other disruptions.

It is not merely plausible that these conditions may have contributed to the negative results reported; there is also factual evidence that the experiments were indeed performed under invalid conditions. Some of the 'control counts' of the two 'negative' experiments were so erratic that they practically varied by 100 per cent! These fluctuations are not mentioned in Maddox's report, and I only realized how large they were when consulting the laboratory book in which these experiments had been recorded. It is on the basis of two experiments performed under conditions hardly compatible with scientific observations that Maddox published a report entitled 'High dilution experiments a delusion'.

Another interesting example of a mock attempt at duplication was provided by a Dutch group in 1992. These investigators studied the effect of high dilutions of aIgE on the staining of basophils in the same range as did the INSERM team (21- to 30-fold dilutions). However, whereas the INSERM scientists had compared the high dilutions of the active chemical (aIgE) with ordinary water or with high dilutions of an inactive chemical (aIgG), the amazing fact is that the Dutch group compared their high dilutions of aIgE with neither of these controls. Instead, they compared high dilutions of aIgE with high dilutions of aIgE, the only difference between the two being that one batch had been violently agitated whilst the second group had been minimally agitated!

The last example of mock duplication was published in December 1993 by a British group. This example is even more interesting than the previous one because, despite the fact that the investigators used a method designed to minimize the chance of actually finding a high dilution effect, they did indeed find a significant difference in the variability of basophils between high dilutions of the active chemical and high dilutions of the solvent. This is a good illustration of how statistical complexity was consequently used to mask an embarrassing result, and is examined in the next section and in technical detail in Appendix 6b.

Finally, it should be noted that, besides the two detection systems described in Chapters 2 and 3, 21 other systems used by 17 different groups have produced positive results so far (Appendix 6d). I conclude therefore that, in 1995, repeatedly claiming that high dilution phenomena cannot be duplicated is closer to an incantation than to a scientific statement.

THE PERVERSE USE OF STATISTICS

In a book entitled *How to Lie with Statistics*, a British scientist quoted Disraeli as having said: 'There are three kinds of lies: straight lies, damned lies, and statistics.' This section describes examples of subtle uses of statistical arguments to dismiss empirical results about homoeopathic dilutions.

One argument quoted *ad nauseam* is that some basophil counts reported by Davenas were considered to be too 'good' to be true. In fact, the statistical argument presented by Maddox and his fraud squad in their report mixed together two different types of situation. The first was that of non-blind duplicate or triplicate counts of 'identical' high dilution samples. Here, repeated counts of the 'same' sample may be influenced by the naïve expectation that the counts should be exactly the same, and lead to an increased clustering of results between repeated counts on identical dilution values. The second situation is quite different, and corresponds to blind counts performed on control samples; these could not be influenced by expectation. Confusing the two types of situation had the effect of *introducing the suspicion of fraud in an implicit way*.

In confusing the two, Maddox and his co-authors considered a deviation from a statistical law called the Poisson distribution to be proof of 'errors' (i.e. incompetence or dishonesty). In fact, however, they did not study the distribution of counts corresponding to the second, blind, situation, but the first, non-blind, situation, which cannot be expected to produce a Poisson distribution. This distribution occurs only when the counts are all independent. In addition, as shown in Appendix 6a, there can be several other reasons (both theoretical and empirical) for deviations from Poisson statistics (e.g. interaction between neighbouring basophils, or effects of staining). Maddox and his fraud squad were so anxious to discount the validity of homoeopathic dilutions that it did not occur to them to consider explanations other than incompetence or dishonesty.

To summarize, their arguments were based on three sorts of confusion:

1. between repetitive counts that were not performed blind and blind counts that could not be influenced by experimenter bias
2. between deviations from a statistical law and fraud
3. between an anomalous phenomenon linked to the staining process of basophils and a phenomenon linked to high dilutions.

Another perverse use of statistics consists in swamping significant results by adding irrelevant data. This kind of strategy has been employed recently by the British team whose failure to reproduce high dilution experiments was published by *Nature* in December 1993. This remarkable publication is analysed in Appendix 6b. To explain its strategy to non-specialists, I will first illustrate it with a fictitious example, in which the readers will know the answer in advance, without needing to perform any statistical calculation.

Imagine a stupid scientist (admittedly an absurd hypothesis) contesting the idea that, throughout history, sons have always been born after their fathers. To demonstrate his thesis, our imaginary scientist could proceed in the following way. Using historical records, he would analyse the birth dates of members of a royal family. After introducing these dates into his computer, he would compare the series of dates corresponding to the fathers to the series of dates corresponding to the sons. In order to achieve this comparison, he would use a standard statistical test. In this test, the difference between two averages is compared to the fluctuations observed within the two series of numbers. In the case outlined here, the two series of dates would appear to be practically identical, for the following reason. Except for the first and the last king, each one appears both as father and as son. Our scientist could then conclude his statistical analysis by writing:

'We have examined two series of birth dates that are particularly well documented, extending over a long period of time. Within the limits of statistical uncertainties, these historical data are compatible with the null hypothesis and furnish no support for the idea that fathers have always been born before their sons. The bizarre belief in this strange theory can probably be explained by the irrational ideas developed by popular media.'

Lest the reader should think that I am putting up a straw man, I reproduce a drawing that appears on the first page of the article published by *Nature* under the title: 'Human basophil degranulation is not triggered by very dilute antiserum against IgE' (Figure 6.1). In the first figure of their article, the authors lumped together experimental data and control data. They also lumped together three series of dilutions performed with *different* blood samples.[3]

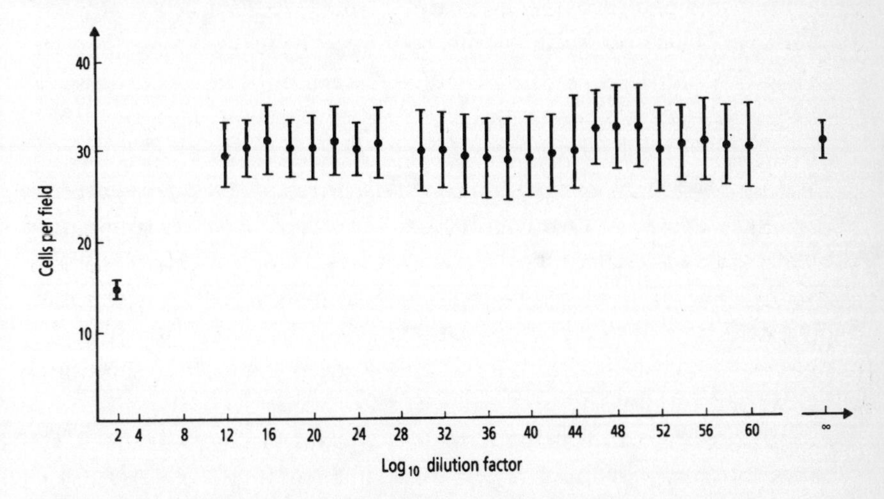

FIGURE 6.1 Figure 1 from the article of Hirst *et al.* (1993), in which the authors combined two sets of control data with their experimental data, thus 'washing out' any possible high dilution effect.

Scientific publications can leave deep scars. A couple of days before the article appeared in *Nature*, the head of INSERM called Benveniste to inform him of this 'failure' to reproduce his high dilution experiments on the staining of basophils. I happened to be talking to Benveniste when the telephone rang and I overheard the conversation. When Benveniste, who had already heard of this 'bad news', tried to explain some of the gross errors contained in the *Nature* report of the failure to reproduce his experiments, the director of INSERM said he had no time to listen to his explanations.

SCIENTIFIC HARASSMENT

As mentioned in the previous chapter, aggressive comments about the work of colleagues are part of the 'normal' behaviour of scientists. However, in normal science, when a scientist is being attacked he is given a fair chance to fight back. In cases of scientific heresy, however, the scales are completely tipped against the unconventional scientist. In order to be allowed to present his case, he is required to reach total perfection. His detractors, on the other hand, are free, nay encouraged, to use any means they wish provided they furnish ammunition against heresy.

Faced with such a bombardment of critical comments, the heretic finds himself in a double bind. He may chose to explore each of the possibilities suggested to him, even if only to show that they are irrelevant; he is then paralysed for years, both by escalating demands for new experiments and new 'controls' and by exhausting battles against editorial censorship. Alternatively, he may chose to ignore his opponents in order to proceed with his research; he will then be accused of being unscientific and deprived of his research tools.

The only technical criticism of basophil experiments published in France provides a good example of what I mean by 'scientific harassment'. The French chemist Jean Jacques published an article suggesting that, in the experiments reported by Davenas *et al.*, the changes in the staining properties of basophils are due not to the 'memory of water' but simply to the agitation used to prepare successive dilutions. The proposal was that oxygen was being dissolved during the shaking of the solutions, which might modify the properties of the chemical used to stain the basophils. In order to test this possibility, Jacques suggested that high dilution experiments should be performed in an inert atmosphere.

The Nobel chemist Jean-Marie Lehn presented Jacques' article to the French Academy of Science and official science gave a sigh of relief. In fact, neither Jacques nor Lehn seem to have read the article of Davenas *et al.* that they had criticized, or at least had not gone beyond the first page. Had they done so, they would have noticed the results of an experiment illustrated in graphical form on page two, in which the effects of two types of high dilutions were presented: aIgE (with positive results) and aIgG (with negative results). Since the two types of solutions had been prepared through the same sequence (diluting by a factor of 10 and shaking), it was impossible to attribute the difference in biological activity to shaking alone.

Faced with the criticism of these chemists, Benveniste did not simply argue that this consideration was irrelevant to his work. He also tried to publish the results of new experiments, which had been performed in a completely blind manner under the supervision of Alfred Spira. When Benveniste wrote Lehn a two-page letter asking him to present this rebuttal to the French Academy of Science, Lehn flatly refused by arguing that he was not competent in this particular field! Benveniste eventually found another member of the Academy to present his article, which was then refereed and published by the *Comptes Rendus de l'Académie des Sciences*.

In a last-minute effort to diminish the importance of this article, a footnote was added stating that it was being printed as a result of the right of response to the article by Jean Jacques. The way this footnote was added illustrates the extent to which scientists go when faced with unwanted knowledge: in order to add this footnote, the whole issue of the *Comptes Rendus* had to be destroyed and printed again.

Encouraged by the editorial published by *Nature* at the same time as the article of Davenas *et al.* of 30 June 1988, scientists responding to this issue in English were much more prolific than those writing in French. Within a few months, seven suggestions were printed by *Nature* concerning hypothetical artefacts that could explain away the unusual results.

As shown in Appendix 6c, each of the suggestions has the same character as the one presented above: it does not account for the experimental results it was supposed to explain. In fact, each of the proposed hypotheses is contradicted by more than one type of observations, the number ranging from two to six depending on the kind of artefact being suggested.

The analysis presented in Appendix 6c illustrates how biased the editors of *Nature* have been in their handling of the battle over the memory of water. For instance they passed for publication a number of negative and derogatory comments by Metzger, which appeared together with his report of the 'failure to reproduce' the controversial results on high dilutions. These included, in a letter to the editor entitled 'Only the smile is left', a description of the 'circus atmosphere' created by results which could be interpreted as supporting homeopathic claims, and of the observations reported by Davenas *et al.* as 'Cheshire cat phenomenon'. Altogether the negative comments of Metzger occupy more space than his actual report of negative results.

IT IS IMPOSSIBLE *A PRIORI,* HENCE IT NEVER HAPPENED

As a final example I will consider one argument presented in the influential report of Maddox, Randi and Stewart entitled 'High dilution experiments a delusion'. At the end of their paper, the investigators mention the hypothetical case of a report concerning a unicorn. Through this argument, they emphasize (quite correctly) that, in order to evaluate the plausibility of a report, one needs to consider the deductive possibility that it may be valid.

For instance, if you are in a variety theatre and you 'see' a woman being beheaded who then rises with a smile, you will probably not reach the conclusion that you have witnessed a miracle. Instead, you will consider what you 'saw' to be the result of a sophisticated trick. This is because, in weighing up the two possibilities, you take account of the fact that in previous observations of beheading no one ever survived such a treatment. On the other hand, there are many past examples of trickery within the framework of a variety theatre. In this particular case, therefore, the *a priori* probability of the beheading being real can legitimately be considered as negligible.

In the case of the memory of water, the situation is completely different because the possibility that the memory of water is a genuine phenomenon is far from being negligible. (*See the theoretical arguments in Chapter 1.*) A scientific debate based on serious consideration of this possibility would represent substantial progress over arguments about unicorns.

Essentially, however, in cautioning against the fallacy of rejecting Benveniste's claims *a priori*, it is not necessary to use technical arguments about the biophysics of water; it is sufficient to ask the following question. An established scientist has made a claim that was almost unanimously rejected by his peers; how many times in the past has such a situation been followed by the acceptance of the controversial claim, and how many times has the controversial claim remained unaccepted by scientists? The *a priori* probability that Benveniste may be right is the ratio of the number of known cases where established scientists dissenting with their peers were eventually vindicated to the total number of known cases of dissent by established scientists. It is of course impossible to evaluate this ratio accurately, but this is not the point, since the number of historical precedents suffices to show that the ratio is not infinitesimally small. Whether this probability ratio is 30 per cent, 70 per cent or only 10 per cent, it is significantly different from zero, so should not be neglected.

It would be equally irrational to consider Benveniste's claims to have been definitively established simply because they have been corroborated by others. The history of science has shown that even an overwhelming consensus can be reversed. For the moment, the only rational attitude is one of tolerance of all the contradictory views concerning the memory of water. This would allow, and hopefully encourage, scientists to explore this controversial phenomenon. So far, however, most technical publications on high dilutions (including the reported 'failures to reproduce') have contained little evidence that could be used to reach a consensus on their reality or otherwise. Instead, what these technical publications have provided so far is a demonstration of the reluctance of many scientists to investigate phenomena that they do not understand.

CHAPTER SEVEN

RUMOURS, SLANDER AND SARCASM

ON 7 September 1992, Benveniste gave a lecture about his research at the Pasteur Institute of Paris. During his talk, he mentioned the theory of the two Italian physicists that predicts that water molecules can organize into 'coherent domains' containing millions of molecules. This higher level organization of water molecules could provide the physical basis for the apparent ability of water to 'remember' previous contact with other molecules. Benveniste seemed to go beyond what his audience could bear when he mentioned new experiments that suggested the possibility of chemical information being transported without any transfer of the corresponding molecules. At that point there was an outburst from one of the scientists. Later on, I learned that this man was a grandson of Jean Perrin, the French physicist who was instrumental at the beginning of this century in making atomic theory respectable in France. Did the man believe that he must defend the memory of his grandfather from an attack against atomic theory? Whatever his motives, he shouted: 'You are insane and you think we are damned fools. We've already had N-rays and mitogenic rays!'

This type of insult breaches one of the strongest scientific taboos, which is that, in principle, all personal attacks are forbidden among academics. Even in their most violent quarrels, scientists are supposed to refrain from such personal attacks and I expected the audience to react to this breach of academic etiquette. In fact, they did not; they were probably relieved that someone should say publicly what many of them thought privately. Retaining his composure, Benveniste replied that he had come to exchange scientific arguments, not insults. He added that psychiatry was a medical specialty, and informed his opponent that he could be sued for illegal practice of medicine.

The point of this anecdote was to illustrate the violent irrationality of some scientists when faced with research which threatens their world view. As we shall see, the history of scientific polemics about the memory of water is full of examples where such insults and other personal slights have replaced technical arguments.

SCIENTIFIC RUMOURS

Rumours are as ancient as humanity. In spite of the fact that scientists pride themselves on only using 'objective' knowledge solidly anchored on 'proofs' and 'facts', they are as prone as the rest of us to use hearsay as a source of information. In fact, they may be even more credulous than lay people because they can be blinded by their methods, naïvely thinking that these give them a privileged access to 'real' knowledge, or even a monopoly over it.

During my very first conversation with another scientist (a French physicist) about the memory of water, I mentioned my intention to study technical publications on the subject and asked him if he would be interested in collaborating with me. In response, he replied: 'They say that Benveniste is mad.' The second colleague to whom I made the same proposal was a US biologist who reacted by saying 'You can't be serious, these guys believe in homoeopathy!' In neither case did the scientist concerned have any personal information on Benveniste or his colleagues.

Rumours violate two of the fundamental rules of the scientific community, that is: (1) reject hearsay evidence and rely only on specific facts that can be discussed in a contradictory debate; (2) never use personal attacks in a scientific argument. However, when science is threatened, these rules can be relaxed. Attacks against Benveniste that have been propagated by rumours have questioned his professional competence, his mental balance and his scientific integrity. I illustrate this fact with some examples of rumours which I have either heard personally or found in scientific publications.

Rumours about Benveniste's professional competence are contradictory. For instance, I have heard people say that, until he started investigating the memory of water, he was one of the few French biologists to be considered as a possible candidate for the Nobel Prize. Conversely, I also heard on two separate occasions that, even before the scandal created by high dilution experiments, he was considered to be a poor scientist. The first time was during an animated discussion, so my opponent may have been carried away in the heat of the debate. However, the second occasion was during a professional seminar and was read as part of a prepared talk. Here are some excerpts from my recording of this talk:

> Immediately after the affair, I tried to make an inquiry by using anonymous questionnaires, where I asked what criteria had been used by biologists and other scientists . . . to have an opinion about the memory of water and it is the only time in my life when I received letters of insults in response to a series of questionnaires. [. . .] How can we understand the fact that Benveniste's reputation among many biologists was so dreadful even before the controversy about the memory of water? Well, one can find a series of indications. One could repeat a series of stories that I have been told which suggest that the controversy about the memory of water finally only confirmed a marginal status that seemed to have been firmly established beforehand. Thanks for your attention.

After this seminar, I tried to gather factual evidence about Benveniste's professional reputation, in particular the prior standing of his publications relative to those of his peers. The simplest way to measure this is to consult the standard tool of bibliographical studies: the *Science Citation Index*. I consequently made a tally of references to Benveniste's publications appearing between 1986 and 1991. I compared this with entries for scientists who also directed an INSERM laboratory and belonged to the same section of INSERM as Benveniste.

In this group of 30 scientists, Benveniste shared the first rank with another INSERM biologist. In 6 years, the number of quotations to his work was above 1000, five times higher than the average of the group. From these data, the scientific reputation of Jacques Benveniste does not appear to be quite so 'dreadful'.

The same positive conclusion about Benveniste's scientific reputation was reached by the editor of the *Science Citation Index*, who published an analysis of Benveniste's publications. In 1989, this editor wrote:

> It is in the interest of the scientific community to note that Benveniste is indeed a highly cited author. A check of the *Science Citation Index* revealed that Benveniste has written dozens of papers, including at least 13 that are cited more than 100 times (Table 2). That is an impressive list. He has written the second most cited paper ever published in the *Comptes Rendus de l'Académie des Sciences*. And certainly a paper from the *Journal of Experimental Medicine* cited more than 640 times is an outstanding achievement. They are both citation classics and we hope he will comment on them in the near future.

Of course, there is no smoke without fire. In this particular case, the fire was probably the sensational article written by Maddox, Randi and Stewart: 'High dilution experiments a delusion', in which the accusations of incompetence remained implicit, so that they contributed to censorship through rumours. The issue of statistical 'errors' has been analysed in the previous chapter. Without referring to this analysis, it can be pointed out that the accusation of statistical incompetence can backfire: if the 'errors' emphasized by Stewart had really been so gross, how can we understand that among the unusually large number of referees who advised *Nature* on the article of Davenas *et al.* no one has noticed these, including Stewart himself? As already pointed out, the low value of the dispersion of repetitive basophil counts appeared explicitly 40 times in the first table of the article.

Because it is so vague, the rumour most easily propagated by hearsay is that of mental imbalance. It is the first one that I met during my investigation. Some 40 years ago, the US physicist Langmuir launched the concept of 'pathological science', in which an established scientist is the victim of 'delusions of grandeur'. This concept became quite popular, so much so that the talk during which it was first developed was reprinted twice in two different journals. The last printing appeared in 1989, during the controversy over the memory of water and over cold fusion. This is probably not a coincidence, because the concept has been utilized by the detractors of the memory of water.

An illustration of the use made of this idea is an article published in *The American Scientist* in 1991 under the title 'Case studies in pathological science', and subtitled 'How the loss of objectivity led to false conclusions in studies of polywater, infinite dilutions and cold fusion'. This article was simply a repetition of the story told by Maddox, Randi, and Stewart, which was further embellished by transforming a 5-day visit into a 3-week one.

As a French example I will mention a book on water published in 1992 by a chemist. In it, the memory of water was included in the list of cases of 'pathological science'. The following year, the same scientist attempted to go a little further. He was planning to quote the memory of water during a conference about pseudo-science as an example of 'scientific delirium'. However, Benveniste saw the conference text as it was circulated as a preprint. He then threatened to sue the author, who finally removed this reference.

Like other rumours, those of fraud are circulated mostly by word of mouth. I was informed of one such rumour in September 1992, when a BBC journalist told me of an interview he had had with a member of the French Academy of Science. This eminent scientist had told him that high dilution experiments were the results of a fraud but had added: 'Don't quote me.'

The suspicion of fraud has also been raised in print. An article entitled 'A threat against science' by another member of the Academy of Science included a section titled 'Reason against dogmatism' in which the author railed first against the Catholic Church and then against the memory of water. According to this author (Evry Schatzman, already quoted in Chapter 5), the memory of water was 'invented' by Benveniste as a means to validate homoeopathy. (This particular example is unrepresentative of the behaviour of scientists because Schatzman is the head of the Rationalist Union, a militant group unrepresentative of scientists as a whole.)

I will limit my other examples intimating fraud to quotes from journals whose articles are listed in the *Science Citation Index*. Because *Nature* has been the most influential source for the suspicion of fraud, I will start by presenting the attitude of John Maddox, its chief editor. Suspicion of fraud against high dilution experiments has been implied in three independent ways:

1 The description of the team visiting Benveniste's laboratory in July 1988 contained the following passage:

> One of us (J. R.) is a professional magician (and also a MacArthur Foundation fellow) whose presence was originally thought desirable in case the remarkable results reported had been produced by trickery. Another of us (W. W. S.) has been chiefly concerned, during the past decade, in studies of errors and inconsistencies in the scientific literature and with the subject of misconduct in science. The third (J. M.) is a journalist with a background in theoretical physics. None of us has first-hand experience in the field of work at INSERM 200.

2 Among the seven experiments that he observed, Maddox disregarded the first four, considering them null and void, including one experiment in which the basophil counts had been made in a blind way.

3 Deviations from the Poisson distribution were presented as being of crucial importance (*see Chapter 6*). Maddox used the issue of basophil counts only as an *implicit* suspicion of fraud. However, he also published an *explicit* accusation based on the same point: in a letter to the editor published on 4 August 1988, a correspondent used the Poisson distribution to argue that the data published by Davenas *et al.* were 'synthetic'.

In order to give the reader a feeling for the way Maddox, Randi and Stewart's investigation was received in certain scientific quarters, it seems appropriate to reproduce the totality of the article published on that report by *The Lancet*, the major British journal of medical research. This article was published anonymously under the title 'Delusion in Clamart':

> 'Hullo, hullo, what's all this?' asked the magician, the editor and the fraud spotter.
> 'It's just a simple experiment,' the professor replied.
> 'You won't mind, then, if we have a look inside your lab, will you, Sir?'
> 'Please do,' said the man from INSERM, a mite apprehensive now at the ordeal his Clamart laboratory was about to undergo.
> So the three visitors to suburban Paris looked at the magic water,[1] returned their eyebrows to the appointed place, blindfolded the assistants with a code stuck to the ceiling, and said, 'Now try.' And nothing happened, which is only to be expected of basophils exposed to water. Orthodox science breathed a sigh of relief. The magician sighed too, at the frailty of human observation. And so did the fraud spotter, though he had found none. And so did the editor, who returned to his tectonic plates, black holes and western blots in Essex Street.
> 'It was not half as much fun as Uri Geller's spoons,' said the magician on his way to the airport.
> 'If anyone else mentions Blondlot's N-rays . . .' complained the editor as he scuttled past.
>
> The supervised test of the degranulating power of infinitely diluted anti-IgE was meant to end with champagne but it closed on July 8 in tears – and with Jacques Benveniste protesting spiritedly at the unreasonableness and disruption of his inquisitors.[2,3] All the same, might there not have been a *frisson* of uncertainty when randomized diluted antibody was first read blind by Dr Elisabeth Davenas and yielded a cyclic pattern of activity even more clear-cut than that reported in *Nature* on June 30?

La Recherche, the French equivalent of *Scientific American*, also contributed to attacks against Benveniste's professional integrity. In this particular case, the mention was not of fraud but only of 'scientific ethics'.

In 1993, *La Recherche* published an article by still another member of the French Academy of Science entitled 'The memory of water, scientific ethics and chance', with the subtitle: 'In 1941, Erwin Heintz "demonstrated" phenomena that resembled the "memory of water". But in 1942, he recognized his error and published an article of retraction.' According to this article, Heintz saw the light and gave up research on the memory of water; in other words, he was honest enough to recant his sins. Allusions to Benveniste were clear, but, to make sure that the connection with Benveniste was inescapable, his name was mentioned three times in the first page of the article. The amusing part of the story is that this 'honest retraction' of Heintz may have been somewhat forced since he still continued to work on high dilutions and even published a long article in 1970 on their effects.

When I first became interested in the memory of water, I had no idea that, some day, I too would be accused of fraud. This accusation was made after I had managed a series of four public blind transmission experiments on 13 May 1993. As shown in the data presented in Chapter 2, the results were not totally positive. After the results were known, someone said humorously: 'As far as faking results goes, there is still room for improvement!' Joking aside, this incident is described in four letters reproduced in Appendix 7a. Briefly, the accusation was made by Charpak, the physicist who had been nominated by INSERM to give his expert opinion on the transmission experiments. During his visit in April 1993, suspicions of fraud had already been aired in a light way, when Charpak told the story of Joliot Curie being shown a 'magic' trick. Here I will only add two comments to those contained in my letter to Charpak. First, I note that Charpak did not answer personally the letter I wrote to him. Secondly, I note that the last word of his letter to Benveniste before the final salutation was the word 'suspicion'.

DEBUNKING AS A SUBSTITUTE FOR SCIENTIFIC ARGUMENTS

As indicated above, I limited my analysis of printed material referring to the article of Davenas *et al.* to articles mentioned in the *Science Citation Index*. These articles are listed in Appendix 7b. In these publications, I found 70 examples of comments that appear out of place in a piece of professional scientific writing. In order to give the reader an idea of the language used, I have included some of the article titles in Table 7.1, in which I have also included examples of text from three cartoons.

TABLE 7.1 *Examples of denigration of the memory of water*

Titles of articles
Case studies in pathological science
Drop of the weak stuff
Homoeopathy – will its theory ever hold water?
Can publishing unbelievable results serve science?
Alternative medicine a cruel hoax – your money and your life?
Citation persective on Jacques Benveniste – dew process at last?
Amadeo Avogadro meets IgE
Delusion in Clamart
When to believe the unbelievable
When to publish pseudo-science
Magic results
'High-dilution' experiments a delusion
Waves caused by extreme dilution
Only the smile is left
Outlandish claims
Benveniste at bay
Tale of the ghostly molecules draws to a close
'Ghost molecules' theory back from the dead
The ghostbusters report from Paris
Science, fringe science and pseudo-science
When water makes scientists shudder
Jacques Benveniste makes an assault on the Pasteur Institute
Benveniste criticism is diluted
A brief history of dubious science
Believing the unbelievable
When the canons of science take French leave

Text from cartoons
Science Faculty Bar: It's the 'dry martini Benveniste'. Just the memory of vermouth
'Molecule memory' man to lose his laboratory: It's our last memory of Dr Benveniste – his tears
Tonight: Jacques Benveniste. Exclusive show of suspense and mysteries. Part one: Joe Baltos and his orchestra

Another example of the use of sarcasm as a substitute for rational arguments against a scientist's unpopular views was a ceremony during which a mock Nobel Prize was attributed to various scientists, including Benveniste. This parody occurred at the Massachussetts Institute of Technology (MIT) in 1991.

In my own writings, I don't hesitate to use irony, but I try to use it as a prop, not as a substitute for rational arguments. However, irony is used by those opposing minority views on the memory of water as a means of closing a debate that has not even been opened. This use of such linguistic forms to suppress a scientific debate on the memory of water is particularly striking when theoretical arguments are barely alluded to, as if they were too obvious to be require any explanation.

In a humorous novel, a French author describes two villages whose inhabitants are not on speaking terms. When someone tries to inquire about the origin of the conflict, a man answers: 'I don't know, but it must be tremendously important because we were already enemies when my grandfather was a child!' As the following list of 'arguments' shows, practically all theoretical objections to the memory of water that have appeared in print are based on the 'logic' that the memory of water is a crime against science that is too enormous to be entertained.

Phrases used by various authors to describe the memory of water include: a 'bizarre new theory', a 'unicorn in a back yard', a 'Catch-22 situation', 'some form of energy hitherto unknown in physics', 'cloud-cuckoo-land', 'unbelievable research results', 'sticking to old paradigms', 'defying the rules of physics', a 'hypothesis as unnecessary as it is fanciful', 'data that did not seem to make sense', 'discouraging fantasy', 'unbelievable circumstance', 'circus atmosphere', 'spurious science', 'magical properties of attenuated solutions', 'unbelievable results', the 'product of careless enthusiasm', a '200-year-old brand of medicine that most Western physicians consider to be harmless quackery at best', 'dilutions of grandeur', the 'egotism and folly of this man who rushes into print with a claim so staggering that if true it would revolutionize physics and medicine', 'mystical powers', 'magic', 'quackery', 'charlatanism', a 'therapy without scientific rationale', 'unicorns revisited',[4] an 'explanation beloved of modern homoeopaths', a 'circus atmosphere', 'spurious science', 'belief in the magical properties of attenuated solutions', 'what seems to be an aberration', 'results that could not be explained by current theory', 'respectful disbelief of Nobel prizewinner Jean-Marie Lehn', the 'cavalier interpretation of results made by Benveniste', 'interpretations out of proportion with the facts', 'magic results', 'high-dilution experiments and much of homoeopathy with their notions of alchemy', 'revolutionary nature of this finding', 'generally efficient physicochemical laws being broken', 'throwing away our intellectual heritage', 'how James Bond could distinguish Martinis that have been shaken or stirred', a 'delusion about the interpretation of the data', the 'extraordinary claims made in the interpretation', 'Cheshire cat phenomena', 'no basis for concluding that the chemical data accumulated over two centuries are in error', the 'circus atmosphere engendered by the publication of the original paper', the 'fact that it still takes a full teaspoon of sugar to sweeten our tea', 'existing scientific paradigms', 'throwing away the Law of Mass Action or Avogadro's number', 'original research requiring a general science background sufficient to recognize nonsense', 'reports of unicorns needing to be checked with particular care', 'not believing that no-more existent molecules can leave an imprint in water', 'the first issue of *New Approaches to Truly Unbelievable and Ridiculous Enigmas*', 'speculating why water can remember something on some occasions and forget it on others', 'outlandish claims', 'not publishing papers dealing with nonsense theories', 'data grossly conflicting with vast amounts of earlier well-documented and easily replicated data', 'extraordinary claims', 'shattering the laws of chemistry', 'divine intervention being probably about as likely', 'findings that contravene the physicochemical laws known to science', 'data that purport to contravene a couple of centuries of chemical data', a 'whole load of crap', '10^{74} oceans like those of the Earth needed to contain only one molecule of the original substance', the 'usual rules of interactions in biology or in physical chemistry where the molecule is the basic vector of information', the 'failure of fundamental principles', 'defying all laws of physical chemistry and of biology', 'unbelievable results', 'observations without any objective basis', one prominent scientist pointedly not reading Benveniste's paper because it would be 'a waste of his time', 'standard theory offering no explanation for such a result' and 'a priest stating during mass that water keeps the memory of God'.

In this avalanche of theoretical objections published in professional reviews, the level of argument appears somewhat low. Amongst all the publications quoting the article of Davenas *et al.*, the *only* reference to a specific scientific law concerns the Law of Mass Action, which governs chemical reactions. Obviously, this law has been checked only for measurable quantities of the substances reacting. There is no reason why a law that has been found to hold within a certain range of values should be true for a different range. Invoking the Law of Mass Action to refute the very possibility of high dilution effects has no more logical basis than invoking 200 years of verifications of Newton's laws of motion to reject Einstein's laws. In both cases, the new phenomena make a difference only in hitherto unexplored regions.

As a last illustration of circular reasoning used against the very possibility of a memory of water, one might mention an argument quoted by Maddox in an article entitled 'Waves caused by extreme dilution', which was supposed to put a final stop to the controversy. The editor of *Nature* wrote that, even if all the matter of the Universe were converted into water, it would not suffice to manufacture the high dilutions presented by Benveniste. This type of argument presupposes that these dilutions could operate only if they contained some molecules of the original product, which is precisely the hypothesis that the new experiments seem to contradict. In fact, however, homoeopathic dilutions are not prepared by adding an 'infinite' quantity of water to active molecules but by successive dilutions into a constant quantity of water.

As I mentioned in Chapter 1, no theory of the memory of water as yet exists. A few authors, however, have provided possible theoretical frameworks for the only question which the detractors of the memory of water have apparently failed to pose, at least in a scientific publication, that is: 'How can one understand the fact that water molecules might possess a structure capable of resisting thermal agitation?' As I indicated in Chapter 1, the theory of coherent domains provides one possible answer to this theoretical question. Another was raised by an author noting the enormous amount of information that could be associated with the fact that both hydrogen and oxygen (the two components of water) have natural isotopes.

To conclude, I would like to give a personal testimony of the attitude of 'I don't want to hear about it' when Benveniste tried to provide a theoretical framework for his research. As I mentioned before, INSERM sent experts in April 1993 to evaluate Benveniste's research. On that occasion, Benveniste was explicitly forbidden to invite Preparata or Del Giudice to explain the possible connection between their theory of coherent domains and the memory of water. Benveniste thus had to rely on a former physicist (myself) to argue with a Nobel laureate in physics (Charpak). In response to my talk, Charpak made a single objection: he had consulted De Gennes (another French Nobel physicist) who had told him that Nozières (a third physicist, who holds a chair at the Collège de France) had assured him that the theory of the Italian physicists was no good. Faced with so irrefutable an argument, I could only keep my mouth shut.

CHAPTER EIGHT

A PSYCHOLOGICAL LOOK AT SCIENTIFIC REPRESSION

So far, scientific censorship has been presented essentially as a manifestation of the reasoning of the strongest within the academic community. However, it is not only a manifestation of strength; like other forms of social oppression, it can also be considered as a manifestation of psychological weakness. One advantage of looking at it in this way is that it can give us strength to fight against it, whereas the former view tends to favour a fatalistic attitude.

Conformity to orthodoxy cannot be explained solely by sociological motives such as professional self-interest. Some of the most violent enemies of the memory of water have no direct professional stakes in the controversy. They have nothing to gain or to lose, one way or the other. When Nobel scientists or members of scientific academies launch attacks against high dilutions, it cannot be because they are hoping for some promotion.

The strong gut reactions against the memory of water led me to believe that there must be personal stakes involved. My assumption is that it is not only the dominant position of a theory, of a discipline or of a scientific leader that is being threatened; the possibility that orthodox science might be wrong in this instance also threatens each individual's world view.

Every one of us has to deal with existential anxiety, and the feeling of belonging to a group is one of the most efficient ways of dealing with it. We derive security from the values of a group, its norms, its rituals and its certitudes. It helps us to construct our own sense of identity. From a psychological point of view, the difference between a group and a sect lies in the fact that, within a sect, the distinction between 'them' and 'us' is quasi-metaphysical. Hence for someone belonging to a sect, humanity is divided into two parts: 'we' who know the truth, and the rest of 'them' who either refuse it or are incapable of having access to it.

Because of its religious connotation, the use of the word 'sect' in a discussion about scientists might appear inappropriate. The latter are vehemently opposed to the idea that their scientific world view might constitute a system of beliefs. But it is precisely this illusion of not having beliefs which characterizes the scientific Church. In this respect, the scientific believer is closer to the marxist believer than to the religious one.

At a time when other value systems (in particular religious or political ones) are losing ground, science appears to be the only safe system remaining. Scientists are the priests of this faith, but we are at the same time more or less part of the community of believers. The need to believe in science can be very strong, even amongst those who understand very little about it, because science is seen as a rock of secure knowledge in an uncertain world.

In my opinion, there is a conflict, or at least a tension, between the quest for meaning and the quest for certainty; what you gain on one side, you lose on the other. The British scientist Cyril Smith recently wrote that 'the ability to tolerate uncertainty is a characteristic of being adult'. From this perspective, scientists have often remained surprisingly young.

CONFORMITY AND SELF-CENSORSHIP

The most efficient policing of the mind is the one we apply to our own intelligence. As stated above, I don't agree with a purely sociological explanation of the conservative attitudes of many scientists. While it is true that the desire to keep a job or to be promoted can favour those attitudes, I think that in many cases this conservatism and conformity are mostly due to psychological rather than socioeconomic motives.

It is not only fear for their careers that has prevented thousands of physicists, of chemists and of biologists from searching for information about the memory of water, but also that of being ridiculed and of losing credibility within the 'scientific community'. During my investigation, I even met two scientists who had observed high dilution effects quite independently of Benveniste, but who had preferred to keep quiet about it.

In one case, the scientist had written an article mentioning that he had observed biological effects beyond the usual range of dilutions. The referee demanded that he withdraw that statement from his article. The author complied and his article was published. Afterwards, he did not pursue the matter any further. In the other case, the scientist had worked for 2 years within a pharmaceutical company on some undesirable effect of high dilutions. After this time, she returned to the CNRS, of which she was a member. In spite of the fact that she was quite convinced of the reality of what she had studied and that scientists belonging to the CNRS have tenure, she did not resume her research . In both cases, the scientists themselves told me their stories spontaneously.

Another example of the role played by the fear of ridicule concerns a scientist who directs an INSERM laboratory. This scientist had been courageous enough to permit me to come with two other people to perform some tests of transmission experiments. Because of a last-minute change in the date of the test, the day that we came coincided with the presence of other outside scientists. I was surprised to hear the director tell us: 'If any one asks you, don't mention that you are working with Benveniste, just say that you are working with me.'

A more extreme fear of being ridiculed was demonstrated to me on other occasions, and was also related to the transmission experiments. One day, Benveniste told a biologist who has important responsibilities within INSERM about a test that I was going to perform in a laboratory close to his. The biologist promised that he would come and observe the test. Actually, he stayed for only 2 minutes, because he was 'too busy'.

On another occasion, in May 1993, I managed a series of public transmission experiments. Benveniste had asked many colleagues to come and watch this demonstration. Most of them did not come, including the physicist who had been nominated as an expert by INSERM to give his opinion on transmission experiments. I also noted that the only person not working with Benveniste who stayed the whole morning was not a research worker but an engineer. It is of course difficult to know to what extent this reluctance of scientists was due to the fear of compromising themselves with Benveniste and to what extent it was due to self-protection of their own belief systems.

In an attempt to break the conspiracy of silence surrounding transmission experiments, Benveniste planned to offer certain scientists an opportunity to perform a transmission experiment. This is why I wrote to a physicist I knew who had publicly assumed a hostile position against the memory of water. In spite of the fact that we had been on friendly terms for the past 20 years, he did not answer my letter of invitation containing the protocol of a demonstration experiment .

To conclude, I wish to emphasize that all the examples above, except for the first two, involved well-established scientists. I think that their reluctance even to come near transmission experiments was due to self-censorship and alienation more than to direct outside censorship. There is no worse censorship than the one we have internalized.

THE ILLUSION OF OBJECTIVITY

The faith in objectivity has only one credo: it is possible to acquire knowledge about the world in a manner that is independent of the person acquiring that knowledge. For those who aspire to objectivity, the scientific method appears to be the grand way to knowledge. One of the leaders of positivist thinking wrote about 'knowledge without knower'; according to this vision as proposed by the philosopher Karl Popper, knowledge exists independently and is waiting to be 'photographed' though the objective of scientific methodology.

In my view, scientific knowledge is the result of an *encounter* between natural phenomena, which exist independently of us, and ourselves, who interpret them. It is not so much the phenomena themselves that are dependent on the observer but the knowledge that we can acquire about them. Scientists are extremely naïve about their own subjectivity; I believe this naïvety to be a major obstacle to scientific knowledge. When scientists do mention subjectivity as an obstacle to knowledge, it is usually the subjectivity of others rather than their own.

According to Karl Popper, the true scientist is someone who tries to contradict his own assumptions. Fighting against his personal beliefs, this ideal scientist attempts to use his observations as ammunition against his own tentative explanation, or hypothesis. After having tried and exhausted all possible ways of refuting this hypothesis, in the absence of contradictory evidence he then temporarily gives it the status of a theory. He still remains willing to reconsider the status of this theory, however, should a new piece of empirical information arise.

From a purely scientific point of view, many authors have emphasized that a one-to-one correspondence between 'facts' and theories is in practice impossible to achieve. Many arbitrary choices are in fact made in the process. What I want to stress here is the psychological naïvety of a vision of science which ignores the crucial role of human subjectivity in scientific knowledge. In some case, this naïvety was quite spectacular. For instance in the *Nature* article that was intended to close the debate in 1988, Maddox mentioned 'true Popperian spirit'. However, an example of his own lack of true Popperian spirit is the fact that, in a television interview broadcast on 5 July 1994, he stated that, the more Benveniste continues to insist on the validity of his experiments on high dilutions, the more discredited he will be. This appears to me a good example of someone who rigidly maintains his position and entirely dismisses the possibility that observations may some day force him to change his mind.

Another illustration of psychological naïvety about lack of objectivity is the behaviour of Metzger. As described earlier, he performed a hasty experiment that was published by *Nature* under the title 'Only the smile is left'. That experiment was supposed to show that the high dilution experiments of Benveniste and his colleagues could not be duplicated. The report invoked a 'fundamental principle of objectivity' to discard the testimony of Italian and Israeli scientists who had collaborated with Benveniste and had succeeded in duplicating some of his high dilution experiments. According to Metzger, these scientists were not 'independent'. However, Metzger himself could be disqualified for not being independent, as he had been a referee of the controversial article and a strong opponent of its publication by *Nature*. In a controversy over attempts to duplicate high dilution experiments, Metzger can therefore be considered to be personally implicated. The fact that this implication could be an obstacle to his own objectivity does not seem to have occurred to him. In other words, 'All scientists are objective but some scientists are more objective than others.'

The myth of scientific objectivity can be considered from various perspectives: historical, epistemological and psychological. Historically, the cult of Reason as a divinity developed in France after the Revolution as part of a movement to eradicate the influence of the Church. Two hundred years later, the battles against religion and 'false' science are being led by the same groups of people. This historical fact poses an interesting problem: scientific fundamentalism is particularly strong in France; how can we understand that France is also the country where homoeopathic medicine is most developed? It can hardly be because Hahneman, the German founder of homoeopathy, took refuge in Paris two centuries ago. In my opinion, the fact that alternative forms of medicine are flourishing in the homeland of cartesian dogmatism constitutes only an apparent paradox. These types of medicine have developed in reaction to the rigidity and inhumanity of official medicine. To the extent that they developed *against* something, they possess their own forms of rigidity and their own brand of dogmatism.

From an epistemological point of view, scientific positivism would have us reject any information that comes from our intuitive perception of reality or from our feelings. I believe this to be unwise. For instance, I became interested in the memory of water partly because I allowed myself to be aware of my feelings of indignation. The fact that I dared to rely on my intuition has also been useful to my investigation. As an example, I can think of my research into the question of basophil counts and Poisson statistics outlined in Appendix 6b. Following my intuition, I tended to trust the people I had met in Clamart, including Elisabeth Davenas. This helped me to go beyond my first impression that her knowledge of statistics was far from perfect.

There are several historical precedents in which scientific progress was initiated by someone relying on his intuition to pursue research that contradicted official orthodoxy. One such example is in the field of human genetics. A couple of albinos had produced a normal child. Albinism is known to be caused by a recessive gene (i.e. one that manifests visibly only where the genetic deficiency is present in both genes). Since the defect was therefore present twice in each of the two parents, genetics predicted that all their children should invariably possess the same genetic defect. Its implication was clear: the mother's husband was not the biological father of their child. However, one scientist involved was so impressed by the parents' protests to the contrary that he decided to look into the matter more deeply. He was rewarded by the discovery that albinism can be caused by either one of two *different* genetic deficiencies. Each of the two parents turned out to have a different deficiency, causing the child to receive one 'good' gene for each of the characters corresponding to the colouring of the eyes. Since one functional gene suffices for each character, the child was normal, contrary to the prediction.

Another case in which the testimony of lay people contradicted official science and contributed to its progress is the phenomenon of meteorites, analysed by Westrum. Peasants had been reporting the fall of heavenly bodies for ages, but scientists had pooh-poohed these popular reports. Never had such a body been observed to fall in the courtyard of the French Academy of Science during a session. It was the intervention of a lawyer that finally pushed the academicians to investigate the matter seriously; trained to deal with human testimonies, he had been impressed by the precision and concordance of the various testimonies.

The third perspective on the myth of objectivity is the psychological one. The American feminist Susan Bordo has analysed the 'flight to objectivity' that is so characteristic of our cartesian culture, specially among males. Georges Devereux, who was trained both in ethnology and in psychoanalysis, has also analysed the flight to objectivity, in a book entitled *From Anxiety to Method in the Behavioral Sciences*. As I have tried to show, it is not only in the behavioural sciences that a rigid use of scientific methods can become an obstacle to knowledge.

When I reflect upon my own history from a psychological perspective, I wonder if most of my colleagues did not become scientists because they made a certain choice during adolescence, namely: always avoid knowledge that directly concerns the self. In other words, objectivity can be not only a tool for knowledge, but also an alibi for the refusal to look inside. We all tend to say 'I don't want to hear about it' when something appears too upsetting. What may be upsetting in the case of scientists is this self-knowledge. In the affair of the memory of water, the 'principle of objectivity' invoked by Metzger and others has not only served to mystify lay people; scientists have also been the victim of self-mystification.

FAITH IN THE INTELLIGIBILITY AND REGULARITY OF NATURAL PHENOMENA

The intelligibility and regularity of natural phenomena are two items of the scientific catechism. These two properties are also part of our intuitive relationship to the world. For instance, an infant notices that, every time she drops a spoon, it never fails to go to the ground. In scientific language, the young child learns that identical causes produce identical effects. As adults also, we have countless experiences that lead us to believe in the regularity of natural phenomena and our ability to understand them. Anything that appears to contradict this vision of a well-ordered world can be a source of anxiety. For example, even such trivial facts as the weather being 'out of season' can sometimes make us uneasy.

The belief in the regularity of natural phenomena is anchored in our culture but I think that it is particularly rigid in people who have been trained as scientists. In everyday life, we are often confronted with events whose internal logic escapes us, but we adjust to and accept our inability to understand everything we see. Suppose for instance, that the printer of my word processor behaves in an erratic way. Since the internal working of printing machines is not part of my expertise, I can accept the fact that I am unable to discern its cause, and simply buy another printer or have it repaired.

Within his field of research, however, the scientist may be unable to accept what is blurred or uncertain. He may then tend to dismiss any questions that cannot be formulated in a precise manner as being 'metaphysical' (e.g. 'What is the meaning of my work?') Most scientists have a need for security and become quite anxious when faced with something unpredictable, such as human behaviour. In my previous book, I showed that, in research dealing with human beings, scientists generally effectively dehumanize their subjects by using standardized tests and other means of 'controlling' the unavoidable uncertainties associated with life.

There are of course positive aspects to the tendency of scientists to seek absolute understanding. This rejection of uncertainty has been a source of scientific discoveries. Whereas the lay person would accept small anomalies, the scientist finds them unbearable. This tendency is an incentive to discover some consistency behind apparently inconsistent events. However, the other side of the coin is that, when an anomaly is impervious to their attempts to make it fit into the current picture, scientists will tend to isolate it outside the field of their conscious knowledge. The memory of water is one example of such unwanted knowledge. In Chapter 1, I mentioned some other anomalies that also seem to threaten the internal consistency of scientific theories and their explanatory power. The subconscious strategy of scientists faced with apparently insoluble questions is often to ignore them until a solution has begun to emerge.

Scientists who refuse to take into account anything that might contradict their vision of law and order in natural phenomena follow a simple rule: what cannot be understood is impossible. They are obsessed by the fear that something should escape them and one of their key words is 'control'. In politics, when the need for law and order is too strong, it leads to the death of democracy. In science, it leads to the stifling of innovative research.

SCIENCE AND THE CITIZEN: THE GOLDEN RULE OF SCIENTISTS

Certain controversies illustrate one of the fundamental rules of scientists. According to this rule, an absolute barrier exists between the inside and the outside of the scientific community as far as technical knowledge is concerned. The story of biomagnetism (*see Chapter 1*) and the reactions to the scientific study of homoeopathic dilutions both illustrate this rule.

When the scandal about high dilutions broke, the INSERM group had already published two scientific papers on high dilutions. However, these reports on such a startling phenomenon had not been discussed by scientists. The extent of the hostility to Benveniste may be explained by the fact that, by seeming to comfort outside knowledge (i.e. homoeopathy), he appeared a traitor to his peers. To make matters worse, the media played a crucial role in the fight against the conspiracy of silence. It seems to have been the report of a major French newspaper on experiments that were being censored by *Nature* that finally forced this journal to publish the article of Davenas *et al.* As a retaliation against what he considered to be intolerable pressure, Maddox reacted with the unprecedented move of coming to Clamart to pass personal judgement. Five years later, when Benveniste tried to explain why the experiments published by *Nature* in December 1993 were far from being an attempt to duplicate his experiments, *Nature* refused to publish the rebuttal he had written with Spira. It took a television programme on scientific heresy to pressure *Nature* into publishing a very short version of that rebuttal, which appeared 7 months after being sent to *Nature*. This long delay can be contrasted with the almost instantaneous publication of innumerable papers hostile to high dilution experiments in summer 1988.

As an example of the highly emotional reactions provoked by the Benveniste affair, I can mention an interview that I had with one of my former teachers. Many years ago, he had encouraged me in my work on genetics and IQ. Later, in 1991, I sent him the manuscript of my book *L'Homme Occulte.*[1] This book contained a 10-page section about the memory of water. For 2 hours, my former teacher spoke exclusively about that section, with a passion that rendered any discussion impossible. It was obvious to me that I had touched a raw nerve.

One of the outspoken enemies of the memory of water has pointed out, probably quite correctly, that the general public would not have been interested in the memory of some obscure chemical. According to him, the word 'water' produces very archaic responses. My personal opinion is that, in the case of the memory of water, as in other controversies involving life, it is lay people who have responded in the right way by being interested; even from a strictly scientific point of view, water is not a trivial substance. The general public has no formal knowledge of the anomalous properties of water and does not know that it is precisely these anomalous properties which have made life possible on Earth. Most people could not care less about the percentage of water in living cells or about the communication problems of molecules within a cell saturated with water. They know nothing about the effect of oscillating magnetic fields on certain cellular processes and remain unconcerned by the theoretical problems posed by high dilution experiments. What they do know and care about, however, is that water plays an essential role in life processes.

In my opinion, the anxiety of scientists faced with the possibility that water really might have a memory goes beyond the stress produced in a conservative mind by possible changes. They talk of several centuries of research being toppled, but, to a large extent, this is only a rationalization. The main issue lies elsewhere. The scientific stakes in the memory of water are real, but I do not think that they alone could produce reactions quite as extreme as what I have observed over the last few years.

The memory of water has acted on scientists like a red flag. I think that what disturbs scientists is a threat to their own image of themselves and to their relationship to knowledge. They take it for granted that they have an absolute monopoly on knowledge about natural phenomena, including those occurring within the human body. Now, imagine that Benveniste and a few others should happen to be right about the memory of water. The revolution would not simply be scientific, it could also become cultural. Should Benveniste eventually be proved right, then homoeopathic doctors and their clients who used such 'folk remedies' would be vindicated and the scientific authorities who have frequently discounted them would look a little foolish. The idea of so disastrous a situation is enough to make scientists shudder as well as those who believe in them.

Churchill once said that democracy is the worst possible system of government, except for all others. One might think of applying the same dictum to scientific research, that is, as the worst road to knowledge about the world, apart from all others. However, this comparison between science and democracy highlights two important drawbacks to the way the scientific community presently operates.

The first has to do with intolerance to diversity. The history of democracies is a long battle against the idea that there could be one single 'good' way of thinking and behaving. When the outcome of an election is that 99 per cent of the votes favour a single party, the result is considered suspect. On the other hand, like monotheistic religions, science believes in only one Truth. Personally, I do believe that there exists a single real world. However, my study of scientific practice makes me think that scientific unanimity can be as suspect as unanimous votes. The issue of tolerance to diversity, though primarily social and institutional, also has important psychological components; once you begin to tolerate a diversity of opinions about the world, you have to learn to accept uncertainty. This is a difficult task indeed, especially for those who have chosen to become scientists.

The second drawback concerns the balance of power. Power struggles and the temptation to use force to obtain and retain it have plagued mankind since the origin of human societies. In principle, if not always in fact, democratic societies recognize the need to have rules which regulate conflicts. In the scientific community on the other hand, checks and balances are practically non-existent; when a conflict arises between a dominant view and a minority one, there are no real provisions to ensure that the minority view will be heard. In the present conflict about the memory of water, for instance, the editor of a major international scientific journal has acted as judge, as accuser, as advocate, as policeman and as head of the appellate court. It is no wonder that those in the minority try to find checks and balances outside this closed shop.

Even societies that believe in the virtues of the free market try to intervene to prevent any group from reaching an economic monopoly. In the question of money for research, however, very little is done to avoid a monopoly by the dominant orthodoxy. Ordinary citizens and their representatives have no way of appreciating the value of dominant theories about cancer. But they can see that progress on cancer has been very slow, in spite of the inpouring of billions of dollars. Just as genetic diversity represents a natural insurance against all possible situations, scientific diversity rather than monopoly is the only possible guarantee against the investment of tremendous resources in a single direction that may eventually lead to a deadlock.

The US Congress recently investigated the issue of fraud in science. In my opinion, censorship and monopoly are much more important issues, which deserve serious consideration from every citizen. In the modern world, science has become too serious a business to be left solely in the hands of scientists. This is particularly true in medical and biological research. When Watson, one of the discoverers of the double helical structure of DNA, was forced to abandon his post as head of the multi-billion-dollar project on the human genome, it was because of a conflict of interest: he had been investing in stocks in genetic engineering. In my opinion, the whole project should have been checked thoroughly to begin with, because of the risk of a dominant position in biological research eventually leading to a practical monopoly.

SCIENTIFIC ARROGANCE: THE CONTRADICTION BETWEEN KNOWLEDGE AND POWER

Each human group tends to have an ethnocentric view of the world, and our modern society is just as ethnocentric as an isolated Indian tribe. The idea of a self-centred vision of the world is in fact quite old. A traditional Chinese tale tells of the sky as seen by a frog. This frog never left the bottom of its well and it thought that the sky was limited to the circle that it could see through the coping of its well. It did not even suspect that a whole universe might exist outside of its well.

At a time when military satellites were able to render every bush in Baghdad visible during the recent Gulf War, we tend to believe that our vision of the world is global. This belief is particularly strong among academics, but most of them have never left the bottom of their 'academic well'. Of course some have travelled physically around the world, attending international conferences, consulting manuscripts at the University of Harvard or Heidelberg, or studying plant species in Africa, but most have nevertheless carried with them their restricted vision of the world. Because they believe that they alone possess the keys to knowledge, academics are usually incapable of learning from other human beings. If my books finally prove to have contributed to knowledge, it will be thanks to groups located outside the ivory tower who have taught me to look at my relationship to knowledge, helping me to change it.

Our relationship to knowledge in the West is one of domination. For instance the manner in which the Western world has reacted to acupuncture is a case in point. Acupuncture developed empirically in a traditional culture and was used for several thousands of years. Recently scientists have substantiated that acupuncture points are indeed located where tradition had situated them and that they can be measured by changes in the resistivity of the skin. The way in which American scientists became interested in one of the possible effects of acupuncture demonstrates our cultural naïvety. During a visit to China, a famous journalist needed an emergency operation. The Chinese doctors used acupuncture instead of drugs as anaesthesia, and the journalist involved was so impressed by this 'discovery' that he wrote a newspaper report on it. Having been validated by a rational human being (i.e. by someone from the West), the status of acupuncture suddenly changed. However, Western scientists failed to ask themselves some basic questions about other possible sources of knowledge, for instance: 'How did "they" acquire such an accurate knowledge of subtle invisible points on the human body without using scientific methods or instruments?' Scientists thus missed an opportunity to discover what they might have learnt from an encounter with another view of the world without translating it and reducing it immediately to their own. In the United States, in order to ensure that acupuncture would not be used in an 'unscientific' way, the use of needles was restricted to medical doctors, despite the fact that a proper acupuncture training requires many years' study – far longer than most doctors have time for.

There are other examples of the tendency of experts to grasp and attempt to monopolize pieces of knowledge coming from elsewhere while perverting their uses and meaning. The example of ecology is well known; after decades of ridiculing the views of 'uneducated' young people, experts are now attempting to limit discussion on this subject to those with 'documented' qualifications. Another example that is not so well known is that of a technique used in early abortions which avoids the need for anaesthesia. This is usually called Karman's technique, but it was in fact imported by him from continental China. It is notable that Karman was not a medical doctor but a psychologist. Consequently, after losing the battle surrounding abortion *per se*, the medical establishment then monopolized it by making it a medical operation requiring anaesthesia.

To conclude this chapter, let us return to the controversy over the memory of water. In this, as in other similar cases, the most urgent question is not the question that can only be answered by experts, that is: 'Does water really have a memory?' Rather it is the following, wider one: 'Do we accept that a closed group should impose their own opinions about truth on everyone, by whatever means fair or foul?' In the case of the memory of water, the public may feel the issue to be relatively trivial. However, when the stakes concern a major public health issue, as in the case of cancer for instance, the arrogant attitude of scientists should be a matter of concern to everyone.

CONCLUSION

JUST before starting the field study described in this book, I published a 10-page report on the memory of water containing the following conclusion:[1]

> The case of the 'memory of water' probably did not contribute to knowledge about the structure of water. On the other hand, it seems to have provided spectacular evidence for one of the psychological limitations of scientists: when they do not know the explanation of a phenomenon, they refuse *a priori* to admit its existence. It is as though we were incapable of perceiving a phenomenon in the absence of a mental picture compatible with our thought habits. From this point of view, scientists appear to be even more rigid than ordinary people.

One of the main themes of this book is that, in a scientific dispute, human factors and technical factors are very strongly and inextricably linked. Failure to take into account both types of factor will result in an incomplete, biased analysis. After 3 years of enquiry I have become more convinced of the scientific import of the observations concerning both the phenomenon known as the memory of water and the possibility that a chemical signal could be transmitted without its original molecular support. However, I am even more confident about the significance of the so-called Benveniste affair as an illustration of the ways scientists suppress unwanted knowledge, and would like to stress that the most important part of my testimony bears not on the memory of water *per se* but on the refusal of scientists to examine evidence that could shatter their current beliefs. In other words, I did not write this book to praise the memory of water but to bury scientific dogmatism. Perhaps a last analogy with the legal process can serve to emphasize this important point.

In certain criminal or political trials, protests may be upheld when the accused has not benefited from due process of law. In such cases, the focus is not on the accused (guilty or not guilty?) but on society (democratic or not democratic?) Were new evidence to show that the accused was indeed guilty as originally judged, the fact that the original trial was unfair would not be modified. In the case of the memory of water, even if new scientific evidence should eventually prove the observations reported by Benveniste and confirmed by others to be erroneous or interpretable in a trivial fashion, the way the case has been handled so far would still resemble a mistrial rather than a scientific debate.

I personally think that future research will eventually validate Benveniste's claims. What I fear might then happen is that, instead of learning something about the dangers of dogmatism, scientists will continue with their business as usual, for 'the greater glory of Science'. This has happened before, for instance in the case of acupuncture. In the case of the memory of water, I fear that the establishment will claim that the scientific method has shown once more its ability to bring us ever closer to the truth about natural phenomena. It will forget the strange incapacity of scientists to face new phenomena, as if their job were to preserve old ideas instead of producing new knowledge. The goal of my testimony about censorship has been to render this kind of rewriting of scientific history a little more difficult. I hope that some lesson will at last be drawn from all this scientific wastage.

Galileo is traditionally considered to be the father of modern science. By referring to his story, I take the risk of dramatizing the issues related to the memory of water. However, because scientists seem so anaesthetized to the dangers of scientific censorship, it is worth drawing some parallels between contemporary scientific censorship and what happened to Galileo.

The obvious parallel between the two relates to intellectual blindness. In this respect, one obvious example concerns those who refused to look down Galileo's telescope. Note, however, that from an intellectual point of view Galileo's case was not as strong as it appears today, after his ideas have triumphed. For instance, the handling of his instrument was difficult and its theory was essentially unknown. The same might also be true for Benveniste's experiments, but in both cases the same essential attitude remains: the inability of experts to face what they don't understand.

The more serious parallel concerns the use of coercion in intellectual disputes. I noted earlier that, whilst socially the use of force is a sign of power, intellectually it is an indication of weakness. The stereotyped image of Galileo chained in a dungeon by the Inquisition is historically erroneous; equally misleading is the inverse image of a man who provoked the authorities of his time through his own arrogance. Though this second image may be psychologically correct, it masks the main question posed by censorship, that is, should we accept a situation in which a dominant group decides without outside control what is proper material for print and what is proper material for research? The fact that a scientist like Benveniste did not risk going to jail, or even risk losing his tenured position, masks the analogy between the two situations; this relationship becomes clearer if, instead of focusing on the individual person, we focus on his research.

In modern democracies, formal prepublication censorship has generally disappeared. In scientific journals, on the other hand, not only is it the general rule, but it is presented as a necessary good, as 'peer evaluation'. Actually, the permission to print given by scientific referees is the modern equivalent of the *imprimatur* previously given by the Church.

Etymologically, 'to publish' means 'to render public through printing'. Scientists, however, use the word 'publish' in the restricted sense of 'validated by the referees of a scientific journal'. Hence, prepublication censorship is the rule in science. Whereas the lack of religious *imprimatur* (licence to publish) used to render subversive publications impossible or dangerous, the lack of scientific *imprimatur* simply renders such publications innocuous. Thus, scientific studies of homoeopathic dilutions could not be published by mainstream journals until the mid 1980s and are still difficult to publish in 1995. The fact that scientists use the word 'publish' in a special way probably masks the parallel between scientific censorship by the gatekeepers of science and the *imprimatur* of the Catholic Church. It should also be stressed that prepublication censorship is only one of the many forms of scientific censorship.

The long history of scientific dogmatism shows that today's heresy could well become tomorrow's scientific truth. An examination of this history suggests that scepticism about expert opinion might be appropriate, especially when health hazards are at stake that involve phenomena outside present-day official science. For the sake of democracy, of public health, and of scientific research, I believe that ordinary citizens should help scientists and other experts to discover the full merits of democratic diversity.

APPENDIXES

APPENDIX 1 A FORGOTTEN ANOMALY: WATER CAN DISSOLVE GLASS!

Those who cite the memory of water as an example of 'pathological science' sometimes refer to the story of 'polywater'.[1] Most scientists who have heard of the latter now believe that it is just another case of 'pathological science', where a well-known scientist imagines that he has made an important discovery, which later turns out to be nothing but an artefact due to contamination. In short, experiments in the 1960s showed that, when capillaries (very thin tubes) are placed above ultra-pure water, a gel can be observed to have formed in them. It was already known that water can change its physical properties when in the neighbourhood of some surfaces, but never before had such drastic changes been observed. The discovery was made by Russian scientists and therefore was at first greeted with sarcasm. Then British scientists confirmed the observations and everybody rushed to study the new phenomenon.

Some time later, it was discovered that the Russian scientist who had reported the phenomenon had made a mistake in interpreting it; this provided an opportunity to bury the anomaly. In a book devoted to the history of 'polywater',[2] one of the world's foremost experts on the scientific study of water wrote in 1981: 'As far as the scientific community is concerned, polywater is now a dead issue.' I chose to mention this issue because the history of 'polywater' illustrates how difficult it is for scientists to face something which they don't understand. The strategy used in the case of 'polywater' is interesting as a historical precedent since it could be re-used to avoid proper consideration of the various anomalies currently associated with the memory of water.

As Franks[2,3] explains in his book, the reality of the case is more complex than it at first appears. Most scientists who have studied the issue now agree that the liquid known as polywater, far from being pure water, was in fact a silica gel. However, even if agreement has been reached on its nature, nobody has yet succeeded in answering the embarrassing question of how it formed in the first place – that is, how can water possibly dissolve such a large amount of glass? Franks makes the point clearly in his book:

> Nevertheless, polywater is not totally dead, since several of the questions it raised have not yet received satisfactory answers. Some of these questions have, on the face of it, considerable scientific merit. For instance, the indications were that water *vapor* reacts with quartz more readily than does liquid water. Philip Low, of Purdue University suggests that 'in their haste to dissociate themselves from anomalous water, members of the scientific community are ignoring important problems'. For example if (as is now accepted) the properties of polywater are due to high concentrations of siliceous material, how could such large quantities of silica be dissolved from quartz and Pyrex capillaries? Quartz is certainly not sufficiently soluble in water to yield the observed silica concentrations, a point repeatedly made by Deryagin in his rebuttals of charges that silica was responsible for the observed properties of polywater.
>
> [. . .]
>
> Any researcher still courageous enough to invoke the existence of 'modified' water must be prepared to face a good deal of ridicule.

This type of censorship consists in taking advantage of any theoretical or experimental error made by the discoverer of an anomalous phenomenon to reject it altogether. It could be re-used in the case of transmission experiments, for instance if it turned out that the transmission reported by Benveniste could not possibly have an electromagnetic origin. If this were to happen, it still would not settle the question. Far from being solved, the mystery would only become deeper – except for those who consider that everything is always for the best in the best of all possible scientific worlds.

NOTES

1 See, for instance: ROUSSEAU, D.L. 'Case studies in Pathological Science. How the loss of objectivity led to false conclusions in studies of polywater, infinite dilutions and cold fusion', *American Scientist,* vol.80, (1992), pp.54–63.

2 FRANKS, F. *Polywater,* MIT Press, 1981, p.145.

3 Felix Franks was the scientific editor of a seven-volume treatise on water: FRANKS, F. (ed.) *Water, a Comprehensive Treatise,* New York, Plenum Press, 1972–82.

APPENDIX 2 HIGH DILUTION EXPERIMENTS PRESENTED IN CHAPTER 2

Inventory of Experiments

Criteria used

Using the laboratory books of Elisabeth Davenas, I have included all experiments fulfilling the following four criteria. (As a check, I also noted the experiments fulfilling all criteria except the first.)

Criterion 1 The blood used must be sufficiently sensitive.[1] This is because high dilution effects are expected to be weaker than ordinary chemical effects so it is assumed that, if a biological detector is not sensitive enough to detect clearly any active molecules when they are present in large quantities, it is unlikely that it will be able to detect the 'memory' of these molecules.

Criterion 2 The number of decimal dilutions must be at least 18. This provides a good margin of safety for assuming no molecules are present because, after the 12th decimal dilution, there is already less than one molecule per basophil.[2]

Criterion 3 The experimental dilution (X) and the control one (C) must have been prepared in exactly the same manner, with the same number of dilution and agitation sequences. This corresponds to the best possible design,[3] in which the *sole* difference between the dilutions compared lies in the presence or absence of active molecules at the beginning of the sequence.

Criterion 4 The counting of basophils must have been performed blind. This eliminates the risk of any systematic bias due to expectations on the part of the person counting the basophils.

TABLE A2.1 ***Solutions being compared***

Series	*Solutions*
A	Apis/serum 10^{-18} or 10^{-30}
B	Histamine 10^{-36}/water 10^{-36}
C	Lung histamine 10^{-30}/serum10^{-30}

List of experiments

Series A, B and C In these experiments, a homoeopathic dilution of a certain chemical (table 2.1) was used to inhibit the usual effect of aIgE on the staining of basophils. In each of these inhibition experiments one high dilution of the homoeopathic product was compared with the same high dilution of water or of serum.[4]

Series D In these experiments, the comparison was between high dilutions of algE and the same high dilutions of the solvent. The dilution range was between 10^{-21} and 10^{-45}.

Series E The experiments were the same as in D but with a more restricted range of dilutions. The range started at 10^{-21}, 10^{-22} or 10^{-23} and ended at 10^{-30}.

Series F This series is comparable to the previous one (E), except that the coding was done by scientists of another INSERM laboratory who coded the dilutions and kept the code until they had received all the results.

Series G In this series, dilutions within the range 10^{-21} to 10^{-30} were compared for two chemical products: algE and algG. At normal concentrations, the first substance inhibits the staining of basophils while the other has no effect. This difference persisted at high dilutions. The results of this series were published in the *Comptes Rendus de l'Académie des Sciences* (Benveniste *et al.*, 1991). As in the previous series, experiments were supervised by Ducot and Spira. Ducot and Spira were also co-authors of the paper published by Benveniste *et al.*

TABLE A2.2 ***Results of high dilution experiments (test 1)***

A
(+) (+) (+) (+) (+) (+) (+) (-) (+) (+) (+)
B
(+) (+) (-) (+) (-) (+) (+) (-) (+) (+) (-) (+) (-) (+) (+) (+)
C
(+) (+) (+) (-) (+) (+) (+)
D
(19+ 5-)[a] (22+ 3-)
E
(10+ 0-) (5+ 4-) (5+ 4-) (6+ 2-) (8+ 0-) (5+ 3-) (7+ 1-) (4+ 4-)
(7+ 1-)
F
(4+ 6-) (6+ 4-) (5+ 5-) (9+ 1-) (10+ 0-) (7+ 3-) (9+ 1-)
(7+ 3-) (9+ 1-) (4+ 5-)[a] (7+ 3-) (4+ 6-)
G
(10+ 0-) (9+ 1-) (6+ 4-) (6+ 4-) (5+ 4-)[a] (7+ 3-) (9+ 1-)
(7+ 3-) (3+ 7-) (8+ 2-) (9+ 1-) (8+ 2-) (2+ 6-)[b] (6+ 3-)[a]
(6+ 3-)[a] (6+ 4-) (9+ 1-) (4+ 6-)

Experiments rejected because of a lack of sensitivity
(6+ 4-) (9+ 1-)[c] (7+ 3-)[c] (5+ 4-) (4+ 3-)[b] (6+ 3-) (3+ 5-)[a]
(3+ 6-) (8+ 1-) (6+ 3-) (3+ 6-) (4+ 5-) (3+ 5-) (1+ 6-)[a]
(2+ 5-)[a] (1+ 7-)

[a]One null result: the result was neither negative nor positive because the number of basophils was the same for the two dilutions being compared.
[b]Two null results.
[c]The second and third experiment had been rejected because of a spontaneous achromasia.

Series H Four inhibition experiments were conducted with histamine to study thermal effects (*see below, test 4*).

Series R These are experiments of the E and F series that had been rejected because the blood used turned out to be insufficiently sensitive. They illustrate the fact that, when the blood is not sensitive enough, high dilution effects are not detected.

Results

Test 1: comparing basophil counts

For each series, bar the first three, Table A2.2 gives the results of individual tests as two numbers: the number of times that the test gave a positive result and the number of times it was negative. For the first three series, however, no number appears because each experiment concerned a single dilution.

For each series, the statistical summary in Table A2.3 gives the number of positive and negative results, and the probability of obtaining results at least as significant as the ones observed through chance alone.

Rejected experiments For these experiments, the score was 72+ versus 67-. The absence of a positive effect illustrates the relevance of criterion 1.

TABLE A2.3 ***Statistical summary***

Series	*Number of results +*	*Number of results -*	*Probability of results occurring by chance*
A, B, C[a]	27	7	1/2000
D	41	8	1/900 000
E	57	19	1/100 000
F	81	38	1/14 000
G[1]	62	27	1/6000
G[2]	58	28	1/1000
Total	**326**	**127**	

[a]In the A, B and C series, several individual experiments corresponded to a repetition with the same pair of products being tested with different blood samples. When the test was repeated with the same pair of dilutions, I only counted one experiment.

Test 2: comparing variances

Series D to G contain 41 experiments for which the variances can be compared within each experiment (i.e. with the same blood samples). For the D series, each experiment was subdivided into three parts, each containing 8 to 10 dilutions. In the summary shown in Table A2.4, the plus sign indicates that the variance is larger for the aIgE dilutions than for the control one, and vice versa for the minus group.

TABLE A2.4 ***Comparison of variances***

Series	*Variance*	
	+	-
D	6	0
E	7	2
F	8	4
G	12	6
Total	**33**	**12**

These results include one experiment of the F series in which the basophil counts were obviously erratic. Depending on whether or not this erratic experiment is included, the probability of the null hypothesis is either 1.4×10^{-3} or 0.74×10^{-3}. If, instead of simply counting positive and negative results, the distribution of the variances were taken into account, the probability would be even lower.

Test 3: the presence of periodic waves

Test 3 requires a large number of dilutions within the same experiment (about 20). Such a large number was rarely used in blind experiments, which are more cumbersome to carry out than ordinary ones. Within the series of experiments that were conducted blind, only the two experiments of the D series contained enough dilutions to allow a quantitative test of the reality of the dilutions exhibiting a wave structure. In the absence of any wave structure, successive counts will as often as not be on either side of the median value. (I used the mean value as a close approximation to the median value.) Successive counts were compared; when it was above the mean line the result was counted as positive, and when it was situated below the result was counted as negative. Using this convention, the results obtained are shown in Table A2.5.

TABLE A2.5 ***Wave effects***

D_1 (solvent)	+-+++-+---+++-+--+--+-+-+
D_1 (algE)	--+++++----+++++----++-
D_2 (solvent)	-+-------+-----+++-++--++
D_2 (algE)	++----+++++---+++++-----+

Each change of sign corresponds to the curve crossing the mean line. The total number of such changes observed is compatible with an even distribution for the solvent (25 out of 48 possible ones) but is grossly uneven for the algE dilutions (12 out of 48). For the algE samples, the probability of the results being due to chance is less than 1 in 1000. It should be remembered that the correlation between successive counts cannot be attributed to a counting bias or to the order in which the numbers were obtained, since all dilutions had been shuffled by the coding procedure, and were therefore blind.

Test 4: the effect of heat on high dilutions

The experiments of the H series were performed on histamine, a product known to inhibit the action of aIgE on the staining of basophils. The dilution used was the 18th centesimal dilution (18CH; i.e. 18 successive 1 in 100 dilutions) of histamine, corresponding to the 36th decimal dilution. This dilution was added to the series of aIgE dilutions corresponding to the so-called second curve (aIgE dilutions between the 8th and the 13th decimal dilution). In order to obtain an unambiguous binary test, I used the following procedure. First I determined which decimal dilution of aIgE had produced the largest effect. I then noted the number of basophils counted for that dilution of aIgE after one of the following four samples of liquid had been added:

S_1: pure water
S_2: Histamine 18CH boiled *after* having been prepared
S_3: Ordinary 18CH dilution of histamine
S_4: Histamine 18CH prepared with histamine boiled *before* the series of 18 dilutions.

S_1 is expected to have no effect. If the memory of water is erased by heat, the second sample is also expected to have no effect. If the 18CH dilution of histamine is effective, it should inhibit the effect of the aIgE dilution. Since histamine is resistant to heat, the fourth sample should show the same effect as an ordinary high dilution of histamine. The basophil counts obtained in four blind experiments are shown in Table A2.6.

TABLE A2.6 ***The effect of heat***

	aIgE alone	*Samples*			
		$+S_1$	$+S_2$	$+S_3$	$+S_4$
	43	43	44	64	54
	40	38	35	52	55
	49	54	57	74	80
	27	32	31	39	41
Average	**40**	**42**	**42**	**57**	**57**

In each, the ranking of the basophil counts corresponds to prediction, thus confirming the other thermal experiments. Even if we simply consider the ranks rather than the actual values obtained, the probability of the results being due to chance is already less than 1 in 1000.

NOTES

1 The sensitivity criterion for excluding an experiment was defined by Benveniste *et al*. (1991) as follows: 'Achromasia below 40% after incubation of the leucocytes with at least two successive dilutions between $\log_2$ and $\log_4$.' This is the criterion used for all experiments reported here.

2 According to the usual model in which many molecules are needed to act on one cell, even a concentration of one molecule per cell would pose problems as regards effect. Actually, most experiments here used dilutions between the 21st and the 30th decimal dilution (i.e. 9 to 18 dilutions greater than the 12th dilution corresponding to one molecule per basophil).

3 In many other experiments, not reported here, the control was simply the undiluted solvent. The advantage of diluting the solvent into itself is that the effect observed cannot be attributed to the sequence of dilution and agitation *per se,* since the same sequence was used for the two dilutions being compared.

4 Serum is water containing salt. The addition of salt is necessary to prevent cells from bursting.

APPENDIX 3a INDIRECT TRANSMISSION EXPERIMENTS USING HEARTS

The period covered by my report on indirect transmission experiments is from July 1992 to December 1993. The transmission experiments presented are the seven public experiments using ovalbumin[1] that were performed blind, plus the three blind experiments that I performed alone. As explained in the test, I was a witness to all public experiments except the first.

Experimental Design

Most control samples contained ultra-pure water that had not been treated by the machine ('naïve' water). Extra control samples sometimes added included transmitted water (where the source tube[2] was naïve water), water treated by the machine with nothing on the input side, or water placed on the output side with ovalbumin on the input side, but without turning the machine on. In the analysis presented here, all samples have been pooled into a single category labelled 'control'.

The coding of the tubes

The philosophy of the coding was that of a contradictory debate; that is, both the experimenters and the observers were in the same position with respect to controlling the manner in which the tubes were being labelled. As an example, let us consider the case of 10 tubes, containing 1 experimental tube (X) and 9 control tubes (C). As soon as it has been prepared, the experimental tube is placed inside an opaque envelope; a sticker marked 'X' is placed inside the opaque envelope. The 9 control tubes are then placed in identical opaque envelopes, each with an inside sticker marked 'C'. The 10 envelopes are then thoroughly shuffled. Once the shuffling has been sufficient to randomize the relative positions of the envelopes, the outside of each envelope is marked with a number from 1 to 10. Finally, each tube is removed from its envelope and labelled with the number appearing on the outside of the envelope. After the tubes have been tested, each envelope will be opened, indicating the significance of the 10 numbers; each of the outside numbers is associated with an inside label indicating either an experimental or a control tube.

Measuring the biological activity of each tube

The measurements were of two kinds. The first was the variation in coronary flow observed for 15 minutes after the liquid being tested had been introduced into the Langendorff apparatus. During the same time period, other mechanical parameters of the heart (e.g. frequency of the heart beats, force of contraction of the heart muscle, etc.) were being automatically recorded and displayed on a screen. In the analysis presented here, however, only the variation in the heart flow was used, because it is the parameter whose variations are most easily quantified. The index of variation used was the absolute value of the maximum amount of variation, expressed as a percentage of the original flow.

In all experiments, two rules were systematically followed: (1) all tubes of a given experiment were tested with the same hearts; (2) for each experiment, at least two hearts were used, in order to have some check on the consistency of the measurements. A third rule used less systematically was: the sequence of measurements with the second heart was the inverse of that with the first heart (i.e. tubes 10, 9, 8 . . . instead of 1, 2, 3).[3]

The Statistical Analysis of Results

Just as there is no unique 'best' way of performing an experiment, there is no unique 'best' way of performing a statistical analysis. Even the way of expressing the results of a single experiment is far from being uniquely determined by that procedure. In the case of transmission experiments, results could be expressed in a least three ways:

1 A numerical value is given for the biological activity of each tube. Results are then presented as two distributions: that obtained with control tubes and that obtained with experimental tubes. This presentation is used in Table 3.1 on page 42. So as to take into account differences in sensitivity of the biological detector from one heart to the next, the raw results were expressed as percentages of results observed with a standard liquid (in this case ovalbumin at a concentration of 10^{-7} mole/litre). Using this method, the average effect observed for experimental tubes differed significantly from zero; the value was 0.35 ± 0.10.

2 Each tube is classified as either 'active' or 'inactive'. Although this way of expressing results seems to be very simple, it is in fact difficult to define this unambiguously, because the sensitivity of the hearts was extremely variable. Benveniste sometimes succeeded in improving this dichotomy by making use of other mechanical parameters besides the heart flow. I myself did not make use of such a delicate classification.

3 The third way of expressing results seems to me to be methodologically the most robust. For each experiment, measurements are used to define the tube that has been most active; this can be done unambiguously by taking a mean average of all results relative to each tube. The classification of tubes obtained in this manner can then be compared with that expected from chance alone; the probability of choosing an experimental tube as the most active one by chance alone is given by the percentage of experimental tubes out of all the tubes. Depending on the experiment, this fraction varied from 1 out of 10 to 1 out of 3.

The actual results of the 10 experiments expressed according to the third method are given in Table A3.1.

TABLE A3.1 ***Results of indirect transmission experiments***

Experiment	*Identity of most active tube*	*Proportion of tubes belonging to the same type as the most active one*
1	X	1 out of 3
2	X	1 out of 3
3	X	1 out of 3
4	X	1 out of 10
5	C	9 out of 10
6	C	9 out of 10
7	X	1 out of 10
8	C	8 out of 10
9	X	2 out of 8
10	X	2 out of 8

In the table, the third column gives the probability of obtaining the result observed by chance. Before analysing these results quantitatively, it should be noted that, because control tubes were always more numerous than experimental ones, each positive result (where the most active tube was the experimental one) had more statistical weight than a negative one (where the most active tube was a control).

A conservative estimate of the statistical significance of the results can be obtained by the maximum likelihood method, using a simple model. The simplest model for the transmission hypothesis is a linear one, with a single parameter x: the probability that the most active tube should turn out to be the experimental tube is $p(x)= x+p(0)$, where $x+0$ corresponds to the null hypothesis that there are no transmission effects and the results are due to chance. The probability that the most active tube should be one of the controls is $1- p(x)$. For a given value of x and a given experiment, the likelihood ratio (transmission effect/no transmission effect) is $p(x))/\{1-p(x)\}$. By multiplying this likelihood ratio for the 10 experiments, the following likelihood ratios L, as a function of x, are obtained:

x	0	0.1	0.2	0.3	0.4	0.5	0.6
L	1	11	49	134	252	340	313

The null hypothesis corresponds to the situation where the two types of tubes have the same biological activity. Even using a simple linear model, this hypothesis is 340 times less probable than the alternative hypothesis (i.e. that the transmission effects exist). Not only is the effect statistically significant, but it is also quite large: a probability that varies between 10 and 33 per cent (under the null hypothesis) is increased by a value x of about 50 per cent. Because the observed increase in x is so large, it is methodologically robust in terms of systematic bias.

A more accurate value of the likelihood of the null hypothesis can be obtained by applying binomial analysis to each group of experiments defined by the proportion of experimental tubes. The expected probability of a positive result was 0.33 for experiments 1, 2 and 3, 0.1 for experiments 4, 5, 6 and 7 and 0.25 for experiments 8,[4] 9 and 10. For each of these groups, the number of positive results expected has a binomial distribution. For instance, in the second group, the numbers of positive results expected by chance alone are shown in Table A3.2.

For this series, the probability of obtaining at least two positive results by chance alone is 5.2×10^{-2}. For the other two groups tested, the probability that chance alone should produce results as good as the ones observed here are 3.7 x

TABLE A3.2 ***Number of positive results expected by chance (group 2)***

Number of positive results	*Corresponding probability*
0	6541×10^{-4}
1	2916×10^{-4}
2	486×10^{-4}
3	36×10^{-4}
4	1×10^{-4}

10^{-2} and 15.6×10^{-2} respectively. The global probability is the product of the three partial probabilities, that is, 3×10^{-4}.

NOTES

1 Some other chemicals have also been used, in particular endotoxin, but these experiments were not systematic enough to be presented here.

2 The source tubes were those tubes placed on the input coil of the transmission machine.

3 The advantage of this method is that it compensates for effects arising from the order in which the tubes are being tested. Although this order was already randomized by the coding procedure, the procedure added another layer of safety. Unfortunately, it also introduced an additional possibility of error, because at any given time the syringe to be used with the first heart could be confused with that designated to the second.

4 For experiment number 8, the actual probability is slightly lower (0.20). By lumping it with experiments number 9 and 10 where the probability is 0.25, I slightly overestimate the probability of the null hypothesis. Another conservative estimate can be obtained by lumping the first three experiments with the last three, using the value 0.33 for the probability of obtaining a positive result by chance alone. With this conservative approximation, the probability of obtaining at least five positive results out of six is 1.8%, yielding a global probability of 0.001 for the 10 experiments taken together.

APPENDIX 3B DIRECT TRANSMISSION EXPERIMENTS USING HUMAN NEUTROPHILS

In autumn 1994, I participated in two series of experiments performed on human neutrophils. In these experiments, the cellular preparations were directly exposed for 15 minutes to whatever is transmitted by the electronic amplifier (e.g. the solvent alone or the active agent).

Experimental Design of the First Set of Direct Transmission Experiments

The first set was a series of blind experiments performed with 10 different blood samples. Each of the samples was used to study the transfer of the molecular signal from four randomized source tubes (two active tubes and two dummies). Four transmissions were thus performed in each experiment, using two transmission machines twice. Each transmission involved one source tube that had been placed on the input coil and two target cell tubes that had been placed on the output coil. The main comparison was between pairs of target cell tubes that had been exposed to the effect of the 'transmitted' active chemical and pairs that had been exposed to the effect of 'transmitted' solvent. As an additional check, a pair of unexposed cell tubes was placed close to the target cell tubes but outside the output coil.

Experimental Design of the Second Set of Direct Transmission Experiments

In a second set of nine transmissions, all source tubes contained the active chemical, but two shielded tubes were placed next to the unshielded ones on each output coil. The shielding consisted of four thin layers of an alloy designed to stop magnetic fields. Again, pairs of unexposed tubes were used as an additional check.

Measuring the biological activity of each tube

The active product used was phorbol-myristate-acetate (PMA). This chemical stimulates the production of oxygen radicals, which can be detected by changes

TABLE A3.3 ***Comparing the effect of transmitting PMA (underlined bold print) with the effect of transmitting its vehicle (ordinary bold print)***

	First sequence				*Second sequence*			
	1st machine		*2nd machine*		*1st machine*		*2nd machine*	
Source	*Unexposed cells*	*Output no. 1*	*Unexposed cells*	*Output No. 3*	*Unexposed cells*	*Output no. 2*	*Unexposed cells*	*Output no. 4*
1	35/38	**36/36**	28/34	**35/32**	34/33	**39/39**	28/22	**44/38**
2	62/65	**91/101**	65/60	**60/64**	63/65	**62/71**	60/63	**84/79**
3	79/84	**80/84**	90/84	**130/112**	88/80	**124/124**	96/92	**92/97**
4	127/116	**152/174**	124/137	**142/139**	122/137	**127/139**	102/107	**126/127**
5	113/126	**161/172**	120/114	**130/121**	113/115	**114/117**	129/114	**162/163**
6	61/53	**92/90**	63/65	**63/73**	76/46	**75/69**	65/66	**97/99**
7	108/110	**120/111**	115/107	**115/114**	106/110	**164/167**	114/119	**225/213**
8	116/116	**151/151**	104/138	**274/186**	116/90	**131/117**	122/139	**138/121**
9	75/67	**115/117**	64/58	**68/65**	72/55	**83/75**	61/62	**149/114**
10	116/119	**120/116**	114/118	**117/126**	116/117	**138/168**	118/117	**159/158**

in optical density in the presence of cytochrome c, measured with a spectrophotometer. The main comparison was between the optical densities of tubes that had been exposed to PMA transmission (T-PMA) and tubes that had been exposed to the solvent, or vehicle (T-vehicle), or had been shielded (shielded T-PMA).

Raw Data Obtained

The raw data are the optical densities measured on each pair of tubes. They are shown for the first set of experiments in Table A3.3, in which the following convention has been used: the optical densities of unexposed cell tubes are given in roman type, while bold print denotes target cell tubes. In the first series of experiments underlined bold print distinguishes target tubes that have been exposed to the active chemical.

Statistical Analysis of Results

The data reported in Table A3.3 show that neither the number of the machine nor that of the sequence has a significant effect on the optical densities. They do, however, depend significantly on the identity of the target cell tubes. The data from shielding experiments show that the 'transmission' effect disappears when the target tubes have been magnetically shielded.

The results of the transmission experiments can be summarized as follows: out of 29 binary comparisons between 'transmitted' PMA and control target tubes (of which 20 were performed blind), 28 gave higher mean optical densities for the cells exposed to PMA 'transmission'. The probability that chance alone should be responsible for such results is completely negligible.

In addition, the statistical estimates presented above are in fact conservative because they are based on comparisons between *single* pairs of tubes, whereas each experiment actually contained two or three pairs. In the first set of experiments (T-PMA versus T-vehicle), the statistical significance of one experiment is given by the probability that *each* of the two T-PMA pairs should have an optical density above each of the T-vehicle pairs; this probability is ⅙. Actually, the order corresponding to a transmission effect is observed in 9 out of 10 experiments ($p < 0.00001$). In the second set of experiments (unshielded versus shielded T-PMA tubes) the mean optical density of *each* of the three pairs of unshielded tubes is above the mean density of *each* of the three pairs of shielded tubes in all three experiments. The corresponding probability of the null hypothesis is 1/20 for a single experiment and 1/8000 for the three experiments. The statistical weight of each experiment is even higher if one considers each individual tube instead of pairs of tubes.

APPENDIX 4 TWO DOCUMENTS ABOUT CONTAMINATED SERUMS

1. Official Letter of Benveniste to the Head of INSERM (17 November 1992) (the translation is mine)

Re: Possible contamination of injectable physiological serum

Sir,

This is an official letter about results obtained during the past weeks. While using as controls Biosedra injectable physiological serum distributed in glass bottles of 500 ml by the Paris social health and medical service [*Assistance Publique*], we obtained the following very strong reactions on isolated hearts of guinea pigs that had been previously immunized. (1) *Decrease of the coronary flow:* when the animal used as donor is very sensitive to endotoxin, in particular after immunization, the coronary flow can be completely stopped. (2) *Mechanical changes:* the most striking change is the sudden decrease of the force of contraction; this decrease can lead to heart failure. Sometimes these effects were observed with pure serum; sometimes they were observed only after amplification (through diluting a thousand times in water, with or without moderate heating). We tested physiological serum from the USA and from Canada that have no effect and have acquired serums from about 10 countries that we plan to test. We have not yet tested the serum provided by the Central Pharmacy of the Paris Hospitals.

The nature of the reactions observed suggests an endotoxin-like activity, but we cannot be completely affirmative. The Biosedra serum certainly contains no endotoxin in molecular form. Since the endotoxin-like activity disappears after the serum has been heated and also under the effect of an oscillating magnetic field (*laboratoire de magnétisme du CNRS, Meudon-Bellevue)*, transfer of an electromagnetic type is a plausible explanation; this transfer could occur either during the manufacturing process or while the serum is being transported, through amplification of some impurity trapped in the glass. It should be noted that, in a preliminary experiment performed with NaCl provided by Prolabo, we also found a similar kind of activity, which disappeared after heating. I predicted long ago the possibility of such an electromagnetic contamination, but, as you know, this was generally met with silence and hostility.

In spite of the fact that we have more than 20 experiments with similar results and, although our controls seem apparently valid, I cannot be absolutely positive about the reality of the phenomenon or about its origin. Such a contamination would probably be harmless to normal human subjects but could have consequences as yet unknown for subjects that have become sensitive to endotoxin through some concomitant pathology.* Urgent measures therefore seem necessary. The first *ad hoc* measure should be the immediate creation of a committee, in order to evaluate these results and, if need be, their origin and their consequences.

I take this opportunity to remind you of the fact that I have been suggesting for years that a committee of experts should be created around the general theme of electromagnetic transmission of biological information. I very much hope that the facts reported here will turn out not to be confirmed, or that they will turn out to be caused by some artefact, which the experts will help us to find. However, if this should not be the case, past negligence of our research establishment could be justifiably criticized. In spite of the fact that I have periodically alerted the authorities during the past years (and again quite recently) about the reality and the importance of this phenomenon, I have been left by myself, without any financial support (my research budget has been regularly decreased). Besides the safety of the patients, this is one more reason for acting quickly and vigorously, especially since the safety measures are simple.

Please answer this letter rapidly. After a week, should this letter remain unanswered, I would feel obliged to warn both the sanitary authorities and the political ones. In view of the tragic events that now fill the news, you will understand my extreme caution. Of course, I don't need to insist on the fact that the information contained in this letter should remain confidential since it deals with matters that could traumatize the general public. But these facts also imply that an evaluation, followed perhaps by adequate decisions, should rapidly follow scientific innovation, in spite of the probable opinion of some 'experts'.

I thank you for your attention. Sincerely yours,

J. Benveniste

Copy sent to the President of the Scientific Council and to the President of Commission no.5 of INSERM

Included: Copy of laboratory book, two typical experiments, 16/11/92

Note

* Please note that hearts coming from ordinary guinea pigs hardly react to endotoxin, even at ponderable doses, while animals that have been immunized become very sensitive. These are classic results of the literature, as are the depressing effects of endotoxin on heart functioning. The results obtained with our model should perhaps lead to research on sudden infant death, where the conjunction of vaccination and Gram infection could play a predominant role.

2. Letter to the Editor of *The Lancet* (16 February 1993)

Sir,

We have detected an endotoxin lipopolysaccharide (LPS)-like activity (ELA) in French 0.9% saline for i.v. injection, our control while studying the cardiac effect of antigen in isolated perfused hearts from immunized guinea-pigs. We noticed that saline alone had an effect and studied it systematically. Saline, native or diluted 1/1000 in Krebs-Henseleit buffer, or *E. Coli* LPS, 1µg/ml, was infused (10 ml, 2 ml/min) into the aorta. Mechanical changes and coronary flow variations (CFV) occurred 14 days after i.p. injection with 1µg ovalbumin (Sigma) in 0.1 ml alum (Alhydrogel[R]) but not in non-immunized animals. Results were as follows:

1 Responses to LPS showed individual[1-3] and seasonal variations. From 26 Oct. to 18 Nov. 1992 untreated diluted saline, $n = 13$, the same heated (70 °C, 1 h), $n = 12$, and LPS, $n = 6$, induced maximal CFV (increase or decrease, in %, mean s.e.m.) of 31.8±7.5, 5.4±0.8 ($p<0.05$, *t*-test for independent variates) and 34.5± 20.1 respectively. From 23 Nov 1992 to 26 Jan 1993 CFV was 14.4±1 .6, $n=43$; 3.8±0.4, $n=35$ ($p<0.05$), and 17.9±3.7, $n=24$ respectively. A major drop in contractile force, sometimes leading to cardiac arrest, was often noted in parallel with negative CFV; in a few experiments, undiluted saline gave similar results. No other i.v. solutes were examined. Recently commenced assays detected no activity in US and Canadian samples (3.5±0.5 and 3.7±0.5, $n=4$), however. Similar assays are projected for samples from 14 countries.

2 Two wistar rats were injected with 1 x 10^7 live BCG (Institut Pasteur, Paris)[2] and 1µg ovalbumin. At *d*-11, infusion of diluted saline triggered sudden cardiac arrest with -100% and 89% CFV. Heated (70 °C, 1h) saline had no effect.

3 Heating diluted saline suppressed saline ELA (see above), which reappeared in about 3 weeks. LPS (1 µg/ml) was unaffected.

What is the nature of saline ELA? CFV, negative isotropism, arrhythmia (in rats) and dependence on pre-immunization closely mimic LPS. Heart sensitivity to LPS (1µg/ml) often paralleled that of saline. However, saline contained no detectable LPS (Limulus test, sensitivity 0.1 ng/ml). ELA was suppressed by moderate heat, contrasting with LPS heat stability. Heat-killed microorganisms are a likely source in saline for LPS amounts below detection level, yet capable of supporting ELA, either directly (thermolability argues against this) or via an undefined mechanism. Also, other cardiotoxic substances remain to be explored as putative cardiac agonists.

Whatever the nature of ELA, its major *in vitro* cardiac affects must be dealt with, since in man, the consequences of high doses of saline are unclear. While harmless to subjects with a normal immune system, it could have adverse effects in those naturally sensitive to LPS, or mounting an immune response, or with immunodeficiency, cancer or haematological and/or infectious disorders.[1-4] In mice and patients with malignant diseases, sensitivity to LPS is linked to high serum IL-6.[4] This marker could help elucidate the nature and immunologic effects of ELA. While in doubt, all solutes for i.v. use should be checked or heated (70 °C, 1 h), thus suppressing ELA for about 2 weeks. New manufacturing processes should be implemented.

INSERM U200 — Jacques Benveniste, MD
32 rue des Carnets — Mohamed Hedi Litime, DSc
F-92140 Clamart — Jamal AISSA

NOTES

1. Galanos, C., Freudenberg, M. A. and Reurrer, W. 'Galactosamine-induced sensitization to the lethal effects of endotoxin'. *Proc. Natl. Acad. Sci. USA*, vol.76 (1979), pp.5939–43.

2. Snell, R. J. and Parillo, J. E. 'Cardiovascular dysfunction in septic shock'. *Chest*, vol.99 (1991), pp.1000–9.

3. Suter, E., Ullman, G. E. and Hoffman, R. G. 'Sensitivity of mice to endotoxin after vaccination with BCG'. *Proc. Soc. Exp. Biol. Med, vol.99 (1958), pp.167–9.*

4. *Yoshimoto, T.* et al. 'High serum IL-6 level reflects susceptible status of the host to endotoxin and IL-1/tumor necrosis factor'. J. *Immunology*, vol.145 (1992), pp.3596–603.

APPENDIX 5 AN EXAMPLE OF INSTITUTIONAL CENSORSHIP: THE DIRECTOR OF INSERM THREATENS BENVENISTE FOR REPORTING AN EXPERIMENT

In July 1992, Benveniste wrote a succinct report of the first transmission experiment which was performed in a blind fashion by scientists not belonging to his laboratory. This report was sent to the 20 scientists who had either followed his first experiments or whom Benveniste was trying to interest in these experiments. The following is a translation of the report and of the reaction of the head of INSERM.

Translation of the Report

On 9 July 1992, an experiment in electromagnetic transmission was performed blind with four outside scientists; the biological activities transmitted were: (1) ovalbumin and (2) *E. coli* endotoxin. The transfer occurred from sealed phials of 2ml to sealed phials of 2ml containing physiological serum. The recipient phials were then diluted 1 in 1000 and 20ml of the dilutions were put into 50ml tubes. The tubes were coded according to the method proposed by Michel Schiff, so that neither the personnel of our laboratory nor the outside scientists had any knowledge of the significance of the code numbers attributed to each tube. The tubes were tested on 11 and 12 July, using the hearts of two guinea pigs that had been immunized with ovalbumin. The code was broken on 13 July. Results are shown in Table A5.1

The probability that such a result could be due to chance alone is 1 in 4000. Moreover, the differences between active and control tubes (both are pure physiological serum!) are clear cut and reproducible. These differences are also found in the mechanical effects (not shown here).

This experiment clearly shows the transmission of a biological activity through an electronic circuit. It demonstrates in an indisputable manner both the electromagnetic nature of molecular information and the role of water as a magnetic memory 'tape' of this information. It also validates the 50 prior experiments of

TABLE A5.1 ***Results of a transmission experiment with guinea pig hearts***

Code number of tubes	*% variation of coronary flow*		*Decision before breaking the code*	*Meaning of the code*
	Heart A	*Heart B*		
1	50	17	+	Endo
2	55	21	+	Endo
3	75	93	+	Ova
4	0	0	-	H_2O^a
5	-50	-53	+	Ova
6	0	0	-	H_2O
7	0	0	-	H_2O^a
8	0	0	-	H_2O
9	0	0	-	H_2O
10	0	0	-	H_2O^a
11	11	10	-?	H_2O
12	-37	-42	+	Ova

H_2O: original phys. water H_2O:[a] phys. water that received information from H_2O

[a]Note from MS: this was a typing error. In the final report, it appeared as H_2OTr. As mentioned in the text, this was another type of control, which in principle should not be distinguishable from H_2O.

the past month. A second blind experiment will be performed.

Reaction of the General Director of INSERM (translation of his letter of 18 Aug. 1992)

You sent me a circular letter dated 27 July concerning the result of an experiment that I might find noteworthy of my attention.

I would like to point out to you that the enclosed sheet contained obvious typing errors (the indications at the bottom of the table about 'H_2O'). In view of the sensitive nature of your research, of which you are well aware, there also appears to be a surprising lack of appropriate explanations ('the probability that such a result could be due to chance alone is 1 in 4000': what result? what difference between H_2O and H_2OTr?)

I very seriously draw your attention to the pernicious character of the spreading of such 'information'.

Should you persist in this type of behaviour, I would be forced to draw serious consequences from it.

Sincerely yours,

APPENDIX 6A AN EXAMPLE OF THE PERVERSE USE OF STATISTICAL ARGUMENTS

The way in which the *Nature* 'fraud squad' handled the question of the dispersion of basophil counts illustrates their prejudices. After photocopying Elisabeth Davenas' laboratory books, Stewart hastily analysed multiple counts corresponding to identical dilutions. This analysis was published twice by *Nature*[1] and was then often quoted by detractors of the memory of water.

The investigators clearly overlooked some crucial points. In his article entitled 'Waves over extreme dilution',[1] Maddox predicted that, in the future, Benveniste would be 'eliminating unavoidable observer bias by making blind measurements a routine'. It does not seem to have occurred to him that many blind experiments had been performed *before* his visit and that the data collected during these experiments could have provided valuable insights into the question of the degree of variability of basophil counts. The point is that multiple counts of the 'same' sample had not been performed blind, even in blind experiments; the coding served to randomize the various dilutions but not the multiple counts of identical dilutions. In order to randomize the position of multiple counts of identical solutions on the microscope plates, the coding procedure would have had to be even more complex than the one used. In other words, the departure from the Poisson law that was denounced by Stewart revealed nothing but 'unavoidable observer bias', due to a non-blind aspect of the procedure.

Another crucial point that seems to have escaped the fraud squad was the fact that the blind experiments did contain some information on the question of possible deviations from Poisson statistics, *independent of observer bias*. In order to obtain this information, one needed only to examine different high dilutions of the control liquid. When I analysed this type of data (obtained both before and after the visit of July 1988), I found that the dispersions rarely reached values as high as those produced during the visit; they also were seldom as small as the values exhibited by Stewart. This illustrates the absurdity of introducing data into a computer without paying any attention to their origin.

In the blind experiments performed at Clamart, the dispersion of control counts measured on series of high dilutions varied according to the circumstances. In particular, it seems that the fact of making measurements in a blind way for several months without any feedback on the quality of the cellular preparation favoured a drift in the stability of the results. On the other hand, when I considered all the values of dispersion available, I concluded that they were significantly smaller than that which the Poisson distribution would allow. The discrepancy seemed to be significant, especially when one considers the fact that experimental errors *other than observer bias*[2] can only *increase* the variance of basophil counts. I must confess that, faced with this strange result, I had a few moments of panic. After searching for possible explanations for this deviation from Poisson statistics, I found the following ones.

Like other 'laws' that are expressed in mathematical form, statistical laws have a hybrid character. The mathematical form of the laws derives from assumptions about the nature of the phenomena and from rigorous chains of mathematical reasoning. As Einstein pointed out: 'As far as the propositions of mathematics refer to reality, they are not certain; and as far as they are certain, they do not refer to reality.'[3] In the case of a statistical distribution like the Poisson 'law', deviations from the expected distribution are often a sign of experimental errors. However, as in other cases when an anomalous result is obtained, the anomaly could also be due to some real phenomenon. The lack of agreement with previous expectations might mean that some of the assumptions underlying these expectations could be invalid.

One of the assumptions used to obtain the distribution of counts known as Poisson statistics is the fact that counts are all independent. In the case of basophil counts, it is *assumed* that the probability of observing a basophil within the small volume corresponding to the average density of basophils is independent of the presence or absence of another basophil within the same small volume. This assumption is neither a law of nature nor one of mathematics. As long as one does not know the exact sequence of reactions leading to the staining of a basophil, it is difficult to evaluate how valid is the assumption of independence.

Several phenomena could produce deviations from this assumption. The simplest would be a physical interaction between neighbouring basophils, which could be either direct (between basophils) or indirect (via an interaction with red cells). Such red cells have indeed shown long range interactions.[4] If the interaction between basophils was direct, a small repulsion would lead to a minor 'exclusion principle', thus decreasing the expected variance.

A third possible influence on the observed variance of *visible* basophils could be due to the staining process itself. Assuming that the staining of a given basophil changes the immediate surroundings of that basophil (through migration of ions, changes in membrane potential, etc.) this may itself act to prevent the staining of the basophil closest to it. Under these conditions, only half of the total number of basophils would be visible. The variance of the number of *visible* basophils would then be only 50% of the expected value.[5]

The idea that basophils might interact in such a way as to reduce the variance is made plausible by experiments published in 1981,[6] that is, 7 years before the Benveniste affair. In these experiments, the variance observed was lower than that expected from the Poisson distribution. Moreover, the amount of deviation increased as the concentration of basophils was increased.

My purpose is not to provide a definitive explanation of a statistical anomaly but to show that the team led by Maddox did not look very hard for such alternative explanations. As La Palice would have said, the first requirement for finding an answer is to really look for it. In the case of the fraud squad, this lack of curiosity is difficult to reconcile with the fact that they used a deviation from Poisson statistics as if it were conclusive evidence of incompetence, with strong implications of fraud.

NOTES

1 'High dilution experiment a delusion', *Nature,* vol.334, (1988), p.287; 'Waves caused by extreme dilution', *Nature,* vol.335 (1988), p.760.

2 In the absence of observer bias, the counting errors add to the variance, as do variations in the volume used to count basophils.

3 Quoted by Hon, G. 'Towards a typology of experimental errors: an epistemological view', *Stud. Hist. Phil. Sci.*, vol.20 (1989), p.471.

4 Rowland *et al.*, 'A Fröhlich interaction of human erythrocytes', *Physics Letters*, vol.82A (1981), pp.436–8.

5 Assuming that the distribution of the number of basophils follows Poisson statistics, the variance will be equal to the mean. If exactly half of all basophils are invisible, the variance will be divided by 4 and the average by 2. The variance of the number of visible basophils will then be 50% of the mean value of that number (i.e. 50% of the 'expected' value).

6 Gérard *et al.*, 'Le test de dégranulation des basophiles humains (TDBH). Intérêt d'une leucoconcentration et du calcul statistique appliqué au taux de dégranulation, *Pathologie Biologie*, vol.29 (1981), pp.137–42.

APPENDIX 6b AN EXAMPLE OF A MOCK ATTEMPT TO DUPLICATE AN EXPERIMENT

In December 1993, *Nature* published an article by Hirst *et al.* entitled 'Human basophil degranulation is not triggered by very dilute antiserum against IgE'.[1] Except for the word 'not', the title mimicked that of Davenas *et al.*, clearly indicating a failure to reproduce the high dilution experiments reported by the INSERM team. After that article appeared, Benveniste and Spira sent a rebuttal to *Nature,* listing over a dozen points in which the British team had failed to follow the protocol of Davenas *et al.* It took a BBC television programme devoted to the controversy over the memory of water (broadcast on 5 July 1994) for *Nature* to publish a short version of that rebuttal.[2] Here, I will limit my remarks to a few crucial points.

It first seems necessary to outline the experimental procedure presented by Hirst *et al.*, because their published report[1] is a model of obscurity on this important point. The parameters of the experiments reported by the British team were as follows:

1 *The blood samples used:* the authors used a different blood sample for each 'session'. Each session used one of three possible treatments (see 2 below) and one of the three possible dilutions (see 3 below).

2 *The treatment used* (indicated by a letter):
A = succussed aIgE
B = unsuccussed aIgE
C = succussed buffer

3 *The dilution ranges* (indicated by a number):
range 1: 10^{-12} to 10^{-26}
range 2: 10^{-30} to 10^{-44}
range 3: 10^{-46} to 10^{-60}
Actually, the authors used centesimal dilutions, so each dilution range contained only eight different dilutions.

In the text, I have reproduced the figure appearing on the first page of that article (*See Figure 6.1, page 91*). This figure appears to summarize the negative results announced in the title of the article. Its significance (or rather the lack of significance) is buried behind an exceptionally long caption. Only the most meticulous of readers will have gone beyond the first visual impression. Other details of the article point to the anxiety of Hirst *et al.* to discredit their opponents through any possible means.

The second page of the article states that 'Davenas *et al.* reported data as percentage degranulation without showing the unmanipulated cell density data.' In fact, it is Hirst *et al.* who have failed to provide a single piece of raw, unmanipulated data. Davenas *et al.* on the other hand provided 40 values of basophil counts, together with the dispersion recorded in four experiments (in Table 1 of Davenas *et al.*). In their 1991 publication, the INSERM scientists published every single value of the basophil counts. But Hirst *et al.* failed to quote that publication.[3]

In December 1993, Benveniste wrote to his British colleagues, requesting their raw data. They answered by stating: 'We are really only prepared to give our raw data to an independent, professional statistician.' In order to examine what can be learned from the published data alone, I will focus my analysis on the three tests presented in Chapter 2 of the text *(see page 28)*.

Test 1 This test concerns the comparison between the number of basophils that are stained (experimental sample versus control sample). In the case of the data presented by Hirst *et al.*, such a comparison would not be valid even if the authors had provided the necessary raw values since they did not use the same blood samples for the A and C sessions.[4]

Test 2 This test concerns the comparison between variances. Despite the fact that the authors did not use the same blood samples for the A and C sessions, they nevertheless report a significant difference in the dispersions observed between the two types of treatment. If this difference is indeed significant, it would correspond to a memory of the liquid used as buffer. The authors dismiss this significant difference by stating: 'It is an interesting feature of our data but it does not, of course, lend any support to the findings of Davenas *et al.*'

Test 3 This test concerns the wave structure of the number of basophil counts as a function of the number of dilutions. Since the authors used different blood samples for the three dilution ranges (with only eight dilutions in each range), no valid test of a wave pattern can be made.

The fact that the authors did not use the same blood samples for the A and C treatments blurred any potential differences between them. In spite of this they found a significant difference for test 2, but dismissed it without providing any explanation.

One last detail illustrates the extent to which the authors neglected to use experimental conditions providing the best possible test of high dilution effects: on one of the experimental samples basophils were used that could not possibly detect any high dilution effects since they were fully insensitive to aIgE even at the high concentration that they started with![4]

NOTES

1 Hirst *et al.*, *Nature,* vol.366 (1993), pp.525-7.

2 Benveniste and Spira, *Nature*, vol.370 (1994), p.322.

3 Benveniste *et al*. 'L' agitation de solutions hautement diluées n' induit pas d'activité biologique spécifique', *C.R. Académie des Sciences,* vol.312, série 11, (1991), pp.461–6. Articles published by the *Comptes Rendus* are systematically compiled by the *Science Citation Index*. It must be noted that these articles contain substantial summaries in English and that tables and figures have bilingual captions.

4 Notice the difference in sensitivity (abcissa = 2) between the sessions corresponding to experimental dilutions (top graph of Figure 3) and those corresponding to control ones (bottom graph). The session with zero sensitivity of the basophils appears in the top drawing (abcissa = 2, ordinate = 0).

APPENDIX 6c SEVEN EXAMPLES OF SCIENTIFIC HARASSMENT PUBLISHED BY *NATURE*

Between 28 July and 20 October 1988, *Nature* published seven letters to the editor proposing explanations other than that which had been proposed by Davenas *et al.* Irrespective of their *ad hoc* character, I label these suggestions scientific harassment because, in each case, the proposed hypothesis is contradicted by several of the observations that it purports to explain. These publications are as follows:[1] (a) Lasters (70), (b) Danchin (53), (c) Suslick (90), (d) Glick (64), (e) Escribano (56), (f) Schilling (83), (g) Shakib (86).

Publication a tries to account for the 'waves' reported by Davenas *et al.* by referring to the two-dimensional matrix formed by successive pits containing the high dilutions. The assumption being made is that a small fraction of each pit spilled over into its four adjacent neighbours. Under appropriate assumptions, this spilling over could produce a pseudo periodic variation of the actual concentration of the active chemical, with a slow decrease in the height of the crests. This hypothesis presupposes that the pits of the matrix were ordered according to successive dilutions. This was indeed the case when these dilutions were not randomized by coding, but the hypothesis becomes irrelevant for blind experiments. It is also contradicted by the fact that, up to the 60th centesimal dilution, the height of the crests remained constant. The inhibiting effect of a single high dilution of histamine, of Apis mellifica or of lung histamine also contradicts this tentative explanation. The same is true of experiments on the physical parameters enhancing the activity of the dilutions (vigorous shaking) or inhibiting it (heating). Altogether, this hypothesis fails to explain five different kinds of observation.

Publication b tries to account for the fact that vigorous shaking is needed. According to this hypothesis, the shaking would provoke extraction of ions from the walls by the initial product (aIgE). These ions would then be equally active in extracting other ions and in inhibiting the staining of basophils. This hypothesis fails to explain both the existence of waves and the constancy of their height. As far as inhibition experiments are concerned, it is difficult to see how the same ions could both mimic the effect of aIgE and inhibit its action. In this case, the hypothesis is contradicted by three kinds of observation.

Publication c suggests that the shaking provokes a local heating at a very high temperature (5000°C) through cavitation. The agent modifying the staining of basophils in the absence of any aIgE molecule is supposed to be some chemical produced by the effect of this heating on some of the molecules of the buffer (whose concentration is the same for every dilution). This hypothesis fails to account for the fact that, when an inactive product (buffer or aIgG) is treated in exactly the same way as the initial aIgE, the resulting dilutions are not active. It also fails to account for the waves and the constancy of their height. As in the previous hypothesis, it is hard to imagine how the same agent (the by-products of molecules of the buffer) can both mimic and inhibit the action of aIgE. Finally, it is hard to imagine how a hypothetical molecule can be produced under a temperature of 5000°C and yet lose its potency at a temperature as low as 70°C. This particular hypothesis is contradicted by five kinds of observation.

According to publication d, the active agent would be heparin, a molecule contained in the buffer. This 'artefact' would not explain the fact that some active product is needed to start the dilutions, the need for vigorous shaking, the waves, or their constant height. As in the previous cases, it is hard to see how the same agent can both mimic aIgE and inhibit its effects on the staining of basophils. This particular hypothesis is contradicted by five kinds of observation.

According to publication e, the active agent would be a by-product of the buffer, produced by the vigorous shaking. Except for the fact that shaking is needed, this hypothesis is contradicted by the same observations as the previous one (a total of four kinds of observation).

According to publication f, the active agent would be a light by-product of algE escaping the process of dilution by always staying at the top of the liquid, like cork-dust in a liquid. This, of course, presupposes that it is always the top of the liquid that is transferred from one dilution to the next. An additional series of *ad hoc* assumptions are also needed: that histamine, Apis mellifica and lung histamine all undergo the same process as does algE, while keeping their antagonistic effects. Even with these additional assumptions, the puzzle of waves of constant height remains, so the hypothesis is contradicted by two kinds of observation.

According to publication g, the loss of staining property reported by Davenas *et al.* is nothing but a spontaneous phenomenon, unrelated to the presence or absence of algE. It is a fact that the staining properties of basophils sometimes spontaneously vary with time. Spontaneous instability of basophils was one of the criteria used to disqualify an experiment, as was the opposite behaviour (the lack of sufficient response to high doses of algE). Apart from the fact that the criteria used were designed precisely to guard against spontaneous changes, the proposed hypothesis rests on the assumption of a constant correspondence between the nature of the product tested (experimental or control) and the time at which the counting of basophils occurs. This could not happen with blind experiments, which randomized the order in which the counting occurred. The hypothesis of a spontaneous effect occurring according to a given time sequence is also contradicted by the fact that it occurs with high dilutions of algE but not with high dilutions of algG. The existence of waves of constant heights remains equally unexplained. As in the case of previous hypothetical agents, it is hard to imagine how the same spontaneous process could both mimic the effect of algE and inhibit it. It is also hard to imagine how the prior heating of high dilutions could inhibit a change in the basophils that is supposed to occur spontaneously. Altogether, this last hypothesis is contradicted by six types of observation.

To conclude: by answering the call of the editor to look for loopholes in the article of Davenas *et al.*, the authors of the seven publications outlined above had a fast ride to a top-level publication. Nevertheless, they have made no contribution to the understanding of the scientific puzzles presented in that article.

NOTES

1 These are listed in their order of publication. The number in parentheses denotes the order in the list presented in Appendix 7b, under the *Nature* subheading.

TABLE A6.1 ***List of high dilution experiments published between 1985 and 1994 in journals cited by* Science Citation Index**

	Tests Used	*Chemicals*	*Dilution*	*Results*
(I) Experiments using tests other than basophil reactions				
1	Immunological reactions (mice)	Thymulin	4 x 10^{-20}g/mouse	+
2	Hayfever	Grass pollens	10^{-60}	+
3	Induced catalepsy (rats)	Gelsemicum	10^{-60}, 10^{-400}	+
	" " "	Cannabis	10^{-60}, 10^{-400}	+
	" " "	Graphites	10^{-60}, 10^{-400}	+
	" " "	*Agaricus muscarius*	10^{-60}, 10^{-400}	+
4	Effect on macrophages (rats)	Silica	2 x 10^{-19}M	+
5	Immunological reactions (mice)	Thymulin	10^{-20g}/mouse	+
6	Retention in blood (rats)	Arsenic	10^{-26}	+
7	Urinary excretion (rats)	Lead	10^{-10}, 10^{-60}, 10^{-400}	-
8	Activation NK cells	β-endorphin	10^{-17}M	+
9	Immunological reactions (mice)	Interferon αβ	2 x10^{-10} i.u.	+
10	Recovery of bowel movements	Opium	10^{-30}	-
	" " " "	Raphanus	10^{-10}	-
11	LRHL release (rats)	PAF	10^{-15}M	+
12	Fibrositis	R toxicodendron	10^{-12}	+
13	Subcellular enzymes	Seven inhibitory agents	10^{-4}–10^{-60}	-
14	Influenza	Homoeopathic drug	10^{-400}	+
15	DIOS cells	Recombinant IL-1	2.5 x 10^{-19}M	+
16	Peritoneal cells (mice)	Interferon αβ	2 x 10^{-10} i.u.	+
17	IL-1 release by THP-1 cells	PAF	10^{-15}M	+
18	Elimination in humans	Nalidixic acid	10^{-14}	-
	" " "	Atenolol	10^{-14}	-
19	Proton relaxation time	Silica	10^{-18}, 10^{-24}, 10^{-30}	+
20	Monocyte fluorescence	PAF	10^{-17}M	+
21	Changes in healthy humans	Belladonna	10^{-60}	+
22	Immune response (chickens)	Bursin	5 x 10^{-27}g/chicken	+
23	Climbing activity (frogs)	Thyroxine	10^{-30}	+
24	Acute diarrhoea (children)	Homoeopathic drugs	10^{-60}	+
25	Asthma	Homoeopathic drugs	10^{-60}	+
(II) Experiments using basophils				
26	Achromasia (human basophils)	Lung histamine	10^{-30}, 10^{-36}	+
	" " "	Apis mellifica	10^{-18}, 10^{-20}	+
27	Achromasia (human basophils)	aIgE	10^{-30}, 10^{-120}	+
	" " "	Na ionophore monensin	10^{-30}	+
		Ca ionophores A23187	10^{-30}	+
28	Achromasia (human basophils)	aIgE	10^{-10}–10^{-30}	-
29	5HT release (rat basophils)	aIgE	10^{-15}–10^{-36}	-
	" " " "	Serum albumin	10^{-15}–10^{-36}	-
30	Serotonin release (rat basophils)	aIgE	10^{-5}–10^{-30}	-
31	Histamine release (human basophils)	aIgE	10^{-5}–10^{-45}	-
32	Achromasia (human basophils)	aIgE	10^{-21}–10^{-30}	+
	" " "	Apis mellifica	10^{-30}–10^{-40}	+
33	Achromasia (human basophils)	aIgE	10^{-21}–10^{-30}	-
34	Achromasia (human basophils)	Histamine	10^{-18}M, 10^{-20}M, 10^{-32}M	+
35	Achromasia (human basophils)	aIgE	10^{12}–10^{-60}	-

APPENDIX 6d SCIENTIFIC STUDIES OF HIGH DILUTION EFFECTS

As of 1995, no review of scientific articles on high dilution experiments has been published. Table A6.1 summarizes the high dilution experiments that were published between 1985 and 1994 by scientific journals listed in *Science Citation Index*. This excludes journals devoted to the scientific study of homoeopathy.

Because reactions to high dilution experiments have focused on those using basophils, it is worth emphasizing that most results have in fact been obtained with other models, including the first one to be published by a mainstream journal.[36] Using mice, the authors of this article demonstrated biological effects of dilutions containing less than one molecule per mouse (2 x 10^{-25} grams injected in each mouse!) This article has been ignored so far, like most articles on high dilutions that followed. The purpose of the summary presented here is to combat one of the most efficient strategies of censorship: simply ignoring what is embarrassing. It should also be stressed that, when usual concentrations such as 10^6 cells/ml are used, dilutions of molecules below 10^{-15} M correspond to less than one molecule per cell (and therefore to much less than one molecule per receptor). The following points can be made about the publications summarized in Table A6.1:

Experiments using tests other than basophil reactions

1 Positive results were reported in 21 publications (out of 25). These results were obtained on 22 different chemicals (or sets of chemicals) by 17 independent groups.

2 None of the four publications reporting negative results contradicts any of the 21 publications reporting positive results.

Experiments using basophils

3 The positive results reported in four publications were obtained on six different chemicals.

4 All positive results were obtained with a sensitive test developed by the authors (achromasia of human basophils).

5 The most influential 'failure to reproduce' (Maddox *et al.*)[28] was obtained under inappropriate circumstances and only concerned two negative experiments (against 200 previously performed, *see Chapter 2*).

6 In three of the reported 'failures to reproduce',[29,30,31] the authors failed to use the achromasia test but used another test, which is apparently less sensitive. In the specific case of histamine release, Beauvais *et al.* showed that it becomes increasingly less sensitive than the achromasia test as the number of dilutions increases.[37] Moreover, two of the experiments [29,30] did not even use human basophils.

7 Ovelgöne *et al.*[33] failed to compare high dilutions of algE to proper controls (algG or the solvent); instead, they compared strongly agitated dilutions of algE to mildly agitated ones.

8 Hirst *et al.*[35] failed to follow the protocol of Davenas *et al.* on several points that are crucial to the sensitivity of the experiment (*see Chapter 6*). In spite of this lack of sensitivity, Hirst *et al.* explicitly reported a high dilution effect on the variance of basophil counts. Unfortunately, the authors failed to publish their raw data and refused to communicate them.

Conclusion

The above survey of published evidence contradicts the idea that 'high dilution effects are not reproducible'. Whereas the effect of high dilutions has been observed in 21 models others than basophils, none of the alleged failures to reproduce basophil experiments actually describes genuine attempts to reproduce these experiments. This situation indicates that the scientific status of high dilution effects is in need of a thorough re-evaluation.

REFERENCES

1 Bastide, M., Doucet-Jaboeuf, M. and Daurat, V. *Immunology Today*, vol.6 (1985), pp.234–5.

2 Reilly, D. T., Taylor, M. A., McSharry, C. and Aitchison, T. *Lancet*, Oct. 18 (1986), pp.881–5.

3 Sukul, N. C., Bala, S. K. and Bhattacharyya, B. *Psychopharmacology*, vol.89 (1986), pp.338–9.

4 Davenas, E., Poitevin, B. and Benveniste, J. *Europ. J. Pharmacol.*, vol.135 (1987), pp.313–19.

5 Bastide, M. *et al. Int. J. Immunotherapy,* vol.3 (1987), pp.191–200.

6 Cazin, J.C. *et al. Human Toxicol.*, vol.6 (1987), pp.315–20.

7 Fisher, P. , House, I. , Belon, P. and Turner, P. *Human Toxicol.*, vol.6 (1987), pp.321–4.

8 Williamson, S. A., Knight, R. A., Lightman, S. L. and Hobbs, J. R. *Brain Behav. Immunol.*, vol.1 (1987), p.329.

9 Daurat, V., Dorfman, P. and Bastide, M. *Biomed & Pharmacother.,* vol.42 (1988), pp.197–206.

10 Mayaux, M. J. *et al. Lancet,* 5 March (1988), pp.528–9.

11 Junier, M. P. *et al. Endocrinology,* vol.123 (1988), p.72.

12 Fisher, P. *et al. BMJ,* vol.299 (1989), pp.365–6.

13 Petit, C., Belon, P. and Got, R. *Human Toxicol.,* vol.8 (1989), pp.125–9.

14 Ferley, J. P., Zmirou, D., *d'Adhemar,* D. and Balducci, F. *Brit. J. Clin. Pharmac.,*

vol.27 (1989), pp.329–35.
15 Orencole, S. F. and Dinarello, C. A. *Cytokine*, vol.1 (1989), pp.14–22.
16 Carrière, V. and Bastide, M. *Int. J. Immunotherapy*, vol.6 (1990), pp.211–14.
17 Barthelson, R. A., Potter, T. and Valone, F. H. *Cellular Immunology*, vol.125 (1990), pp.142–50.
18 Ferry, N. *et al. Brit. J. Clin. Pharmac.*, vol.32 (1991), pp.39–44.
19 Demangeat, J. L. *et al. J. Med. Nucl. Biophy.* vol.16 (1992), pp.135–45.
20 Rola-Pleszczynski, M. and Stankova, J. *J. Leukocyte Biol.*, vol.51 (1992), p.609.
21 Walach, H. *J. Psychosomatic Research*, vol.37 (1993), pp.851–60.
22 Youbicier-Simo *et al. Int. J. Immunotherapy*, vol.9 (1993), pp.169–80.
23 Endler, P. C. *et al. Vet. Human Toxicol.*, vol.36 (1994), pp.56–9.
24 Jacobs, J. *et al. Pediatrics*, vol.93 (1994), pp.719–25.
25 Reilly, D. *et al. Lancet*, vol.344 (1994), pp.1601–6.
26 Poitevin, B., Davenas, E. and Benveniste, J. *Brit J. Clin. Pharmac.* vol.25 (1988), pp.439–44.
27 Davenas *et al. Nature*, vol.333 (1988), pp.816–18.
28 Maddox, J., Randi, J. and Stewart, W. W., *Nature*, vol.334 (1988), pp.287–90.
29 Metzger, H. and Dreskin, S. C. *Nature*, vol.334 (1988), p.375.
30 Seagrave, J. C. *Nature*, vol.334 (1988), p.559.
31 Bonini, S., Adriani, E. and Balsano, F. *Nature*, vol. 334 (1988), p.559.
32 Benveniste *et al. C. R. Acad. Sci. Paris*, vol.312 II (1991), pp.461–6.
33 Ovelgönne, J. H. , Bol, A. M. J. M., Hop, W. C. J and van Wijk, R. *Experientia*, vol.48 (1992), pp.504–8.
34 Sainte-Laudy and Belon, P. *Agents Actions*, vol.38 (1993), pp.C245–7.
35 Hirst, S. J. *et al. Nature*, vol.366 (1994), pp.525–7.
36 Doucet-Jaboeuf, M. *et al. C. R. Acad. Sci. Paris*, vol.295 (1982), pp.283–6.
37 Beauvais, F. *et al. J. Allergy Clin. Immunol.*, vol.87 (1991), pp.1020–8.

APPENDIX 7a FOUR LETTERS ABOUT A SUSPICION OF FRAUD

Letter of Benveniste to Charpak (14 May 1993)

Sir,

I am rather worried about the way things have evolved. I think that you are aware of how serious the simple use of the word 'fraud' can be. Fraud had never been mentioned, except by irresponsible journalists, who were condemned by the court, for lack of any evidence. I regret your absence during the coded experiment of 13 May. You would have seen that the way it took place showed that every precaution had been taken against the possibility of some system of recognition. The point of the coding was not to combat fraud, which is out of the question for everyone in this laboratory, but simply to avoid any possible bias of the technicians. Note that they receive numbered syringes that have been prepared by another technician, which means that they never see the original tubes.

When we heard of your coming, we said: 'Finally a scientist!' It is therefore quite disappointing to hear that you are taking up again gossip which we thought we had been rid of since 1988. The idea that 'someone is cheating behind Benveniste's back' was the way out used by *Nature's* group with its magician. At the present time, at least 10 people are involved in the experiment; each of them is thus under trial. Usually, scientists choose their best results once they are convinced that their hypothesis has been demonstrated. We do not act that way, but show everything to everybody, thus taking the risk that misinformation of the worst kind might come out of it. Sir, act as a scientist, not as a cop. What we have found, almost by chance, is indeed enormous. The stakes are beyond both of us. Given the issues involved, mediocre attitudes cannot be justified and are intolerable. You don't understand? Neither do I. But it exists. Contribute to the outcome of truth. Otherwise, if progress continues at the present rate, you would at best (or at worse) delay it by a few months, since testing systems are becoming easier to handle and more demonstrative.

Concerning the difficulty you have in understanding what is going on with this machine, you are not the only one. As you well know, the argument: 'I don't understand, therefore it does not exist' has been used so often in the past that it is completely discredited. In the near future, we are planning to perform experiments with a direct coupling between the two coils, in order to find out if amplification is necessary. But I remind you of the fact that when we move the potentiometer a few millimetres, the induced frequency of 1000 Hz disappears, as does the transfer. This is not the behaviour of a piece of wood. And even if it were a piece of wood, the fact remains that, with this piece of wood, we endow water with a complex antigenic activity! Better yet ...

However, the best way to cut short any suspicion of fraud would be for you to perform the experiment yourself in your laboratory. I remind you of the fact that this is what I had initially suggested (instead of Cochin). The experiment would be performed by two outside observers designated by both of us* who would guarantee that the transfers occur according to a protocol that has been defined in advance. I am going away for a 10-day visit to the United States around Ascension Day. When I return, we can determine how to establish a protocol. This experiment should remove all doubts that seem to have persisted for the last 5 years among a few individuals that are apparently endowed with a high contaminating power. I emphasize again that, although we go through these experimental pantomimes because of scientists who, today like in the past, refuse to admit a result that disturbs established theories, this is nevertheless contrary to scientific ethics. The normal attitude of a scientific institution should be to judge a result for itself, rather than according to its scientific consequences; otherwise, one might as well be influenced by economic consequences, or even political ones (Lyssenko). Any other attitude leads to arbitrary decisions and to the sterilization of research, since, otherwise, any deviant result becomes suspect, and hence to be gunned down. Once the *usual* precautions have been taken to ensure its validity (and this is more than the case in the present story) any fact must be published, so that the scientists can validate it or invalidate in the following years. If a fact is to be duplicated *before* being printed, research stops. Today, could Planck escape the referees of *Nature?*

Therefore, following these two blind experiments, I will perform only one other verification: the duplication of the experiment in Coraboeuf's laboratory. Although it is also stretching the usual rules of scientific behaviour, it should allow a reversion to normal scientific discussions. But first, you must make a transfer with your own hands. After all, this is what Eccles did, who found out that Loewi was right and he, Eccles, was wrong. You are a man of honour: you cannot make remarks that are degrading to a colleague and refuse to perform a verification that would stop the rumour.

I thank you for your cooperation. Yours sincerely,

J. Benveniste

* For instance Pr Spira, who is the president of section 10 of INSERM.

Letter of Schiff to Charpak (16 May 1993)

Sir,

Last Friday, I learned from Mr Benveniste that the report made to you by Mr Lewiner (or by Mr Hennion, I don't know which) about the series of 4 demonstrations that I managed on Thursday the 13th of May within the laboratory of Unit 332 of INSERM at Cochin led you to be convinced that this series of demonstrations must have been vitiated by fraud, a fraud of which I was probably the agent. In an affair as complex and as delicate as this one, the fact of going through intermediaries increases the communication problems. This is why I prefer to communicate with you directly.

Since it seems that my behaviour led you to certain assumptions of a psycho-epistemological nature, it is only natural that I should in turn present my own interpretations. You will find enclosed an analysis which I published last year about what I called 'epistemological repression'. Your own behaviour in this affair seems to illustrate what I was writing then:

1. The Director General of INSERM refused to allow Preparata or Del Giudice to come and explain their theory of coherent domains, which, for the moment, seems to me to be the most promising one to solve the epistemological riddle posed by Benveniste's experiments on the memory of water. Because of this refusal, I was led to act as a substitute and to formulate what I thought I had understood of this theory. I sent you a text beforehand and I gave an introductory talk at the beginning of your visit on 21 April 1993 to which you seemed to respond through an argument of authority. I attempted to explain the theory of coherent domains and its potential links with the memory of water, using whatever competence was left to me as a former physicist. You, on the other hand, seemed to me to have reacted through an argument of authority, by explaining that you had consulted Mr De Gennes, who himself had referred to Mr Nozières, who, according to him, had declared that the theory of coherent domains had no validity.
2. After my talk of 21 April, you made an allusion to the possibility of mystification by presenting an anecdote about your past work with Joliot-Curie; after being a witness to a magician's trick presented by colleagues, he asked them: 'Where is the trick?' You will agree with me that the balance of power and the circumstances did not favour a serene discussion on this point. This is why I chose not to raise the issue.
3. I had been told that you were going to be present to watch the demonstrations on 13 May, but you delegated two persons to represent you.
4. It seems that you interpreted my temporary irritation and the fact that I objected to your delegates interfering with an ongoing experiment as indicating that a fraud must have occurred. In case your informers did not report it, I mention the fact that I insisted that they should watch at least one of the four experiments; I also insisted that they should accept to play the role of participant–observer or of witness described in the protocol. They refused and I insisted that they should at least be present to watch one of the experiments. Actually, they spent only half of the duration of one experiment in the demonstration room. What provoked my irritation was the fact that, instead of watching the ongoing operations, they turned their back to the apparatus and proceeded to argue with Benveniste about fraud and about the 'open-mindedness' of the scientific community, which, according to them, is not as narrow-minded as Benveniste claimed. You must admit that I had excuses for losing my temper!

In concrete terms, I propose that you delegate someone from the Paris School of Physics and Chemistry (or from somewhere else) to negotiate with Mr Benveniste and with me a demonstration protocol that could possibly lead to a change in your opinion about the existence or non-existence of a separation between a chemical signal and its original molecular basis. Should your assumption of fraud be unamenable to change, there would of course be no point in trying to modify the protocol; since it could not be refuted in an experimental way, this assumption would appear to be non-scientific, at least according to the definition given by Popper. As I explained in the protocol, such was the stake of the experiments: the one you watched on 21 April and those to which you sent two representatives on 13 May.

I will end this letter with a note that is less epistemological and more personal. As I told you on 21 April, I had been looking forward to a collaboration with my former Alma Mater to clarify a scientific enigma. I must confess that I have been disappointed so far, but I don't give up the hope that reason might triumph at the Paris School of Physics and Chemistry and elsewhere.

Michel Schiff, former student of the School of Physics and Chemistry (class of 1956). Chargé de recherché au CNRS.

Enclosure: Chapter entitled: 'Je ne veux pas le savoir: le refoulement scientifique à l'état pur' (*L'Homme Occulté: le Citoyen face au Scientifique*), 1992, Editions Ouvrières, pages 63–78 ['I don't want to hear about it: scientific repression in its purest form' (*Humanity Masked: the Citizen facing the Scientist*]

Copies to: Jacques Benveniste (INSERM, U200, Clamart); Jacques Lewiner and Claude Hennion (ESPCI, Paris); Emilio Del Giudice and Giuliano Preparata (University of Milan); Isabelle Stengers (Free University of Brussels)

Letter of Lewiner to Schiff (18 May 1993)

Sir,

I received a copy of the letter you sent to Mr Charpak on 16 May and it seems important to dissipate the wrong interpretation that seems to have developed after our visit of 13 May.

Actually, on our return, we communicated to Mr Charpak our feeling about the experimental procedure chosen and we proposed one that differs very slightly and seemed to us susceptible either to convince the scientific community or to show the necessity of additional experiments.

We certainly never claimed that the series of demonstrations that you conducted was vitiated by fraud, and *a fortiori* fraud perpetrated by you.

Therefore, we shall propose to Mr Charpak an experimental procedure which, if it seems to him to be of interest, will probably be submitted to you.

In the meantime, I am sincerely yours,

J. Lewiner

Letter of Charpak to Benveniste (19 May 1993)

Dear Mr Benveniste,

Please excuse my delay in answering your messages. I was not available because of journeys and conferences.

However, I made certain that two of my co-workers of the School of Physics and Chemistry should go to Cochin, because their collaboration is essential for laboratory tests. They confirmed to me that the amplifier oscillated in a permanent way. But after thinking about it, I do not intend to draw any conclusion from it for the moment.

The effect which you observe, and you say that it is easily reproduced, needs only a simple test. The use of about 20 phials, some of which have been sensitized according to your method, using a protocol determined by you and without your being able to know the distribution of the phials, should permit an objective test.

During the visit of my co-workers at Cochin, there was a small discussion with Mr Schiff because they thought that they had noticed a possibility of marking the phials that had been sensitized during the phase of vibration. This certainly does not mean that this possibility was used. But it is clear that no doubt should remain. It will be easy for Mr Spira to define a protocol forbidding any suspicion.

Sincerely yours,

George Charpak

Copies to: Mrs Coraboef, Hennion, Lazar, Lewiner, Spira.

APPENDIX 7b THEMATIC ANALYSIS OF THE ARTICLES QUOTING DAVENAS *ET AL.*

The Corpus Used

I used all the articles in English and in French that I was able to consult in 1993. These articles were located essentially through the *Science Citation Index* and through Garfield's article quoted below as reference 22. The journals are listed in alphabetical order. Then the first authors are given, also in alphabetical order. For further use, each reference is designated by its rank in the list.

American Scientist (Rousseau, vol.80 (1992), pp.54–63)[1]

Biochemical and Biophysical Research Communications (Kahn, vol.166 (1990), pp.1039–46)[2]

Biological Journal of the Linnean Society (Berezin, vol.35 (1988), pp.199–203)[3]

British Journal of Clinical Pharmacology (Ernst, vol.30 (1990), pp.173–4),[4] (Ferley, vol.27 (1989), pp.329–35),[5] (Ferry, vol.32 (1991), pp.39–44)[6]

British Medical Journal (Davies, vol.299 (1989), p.918),[7] (Fisher, vol.297 (1988), p.618),[8] (Grange, vol.300 (1990), p.47),[9] (Smith, vol.297 (1988), pp.377–8),[10] (Wall, vol.299 (1989), p.1401)[11]

Canadian Medical Association Journal (Morgan, vol.146 (1992), pp.1719–20),[12] (Richmond, vol.147 (1992), pp.97–8)[13]

Canadian Veterinary Journal (Scrimgeour, vol.31 (1990), p.336)[14]

Chemical Engineering News (Dagani, vol.66 (1988), p.6)[15]

Comptes Rendus de l'Académie des Sciences (Benveniste, vol.312 (1991), pp.461–6),[16] (Jacques, vol. 310 (1990), pp.1437–9)[17]

Critical Reviews of Oncology/Hematology (Valent, vol.10 (1990), pp.327–52)[18]

Current Contents (Anonymous, no.39 (1988), pp.9–10)[19] (Anonymous, no.13 (1989), pp.8–10),[20] (Beaven, no.21 (1990), pp.6–8),[21] (Garfield, no.13 (1989), pp.3–7),[22] (Garfield, no.21 (1990), pp.3–5)[23]

Dermatology (Aberer, vol.182 (1991), p.253)[24]

European Journal of Pharmacology (Sassard, vol.183 (1990), p.384)[25]

Experientia (Ovelgönne, vol.48 (1992), pp.504–8)[26]

Genetics (Stahl, vol.132 (1992), pp.865–7)[27]

Journal de Chimie Physique (Berg, vol.87 (1990), pp.497–515)[28]

Journal de Médicine Nucléaire et de Biophysique (Demangeon, vol.16 (1992), pp.135–45),[29] (Spira, vol.16 (1992), pp.105–6)[30]

*Journal of Allergy and Clinical Immunology (*Beauvais, vol.87 (1991), pp.1020–8)[31]

Journal of Nuclear Medicine (Scott, vol.33 (1992), pp.407–9)[32]

*Journal of the American Mosquito Control Society (*Mehr, vol.6 (1990), pp.469–76)[33]

Journal of the Royal College of General Practitioners (Swayne, vol.39 (1989), pp.503–6)[34]

Journal of the Royal Society of Medicine (Lewith, vol.83 (1990), pp.543–4)[35]

Lancet (Anonymous, 9 Jul. (1988), p.117),[36] (Anonymous, 6 Aug. (1988), p.347),[37] (Benveniste, 13 Oct. (1990), p.944)[38]

Medical Hypotheses (Berezin, vol.31 (1990), pp.43–5)[39]

Nature (Anonymous, vol.333 (1988), p.787),[40] (Anonymous, vol.334 (1988), p.367),[41] (Anonymous, vol.340 (1989), p.82),[42] (Benveniste, vol.334 (1988), p.291),[43] (Benveniste, vol.335 (1988), p.759),[44] (Bland, vol.335 (1988), p.109),[45] (Bonini, vol.334 (1988), p.559),[46] (Claire, vol.335 (1988), p.584),[47] (Clemens, vol.335 (1988), p.292),[48] (Coles, vol.334 (1988), p.372),[49] (Coles, vol.340 (1989), p.89),[50] (Coles, vol.340 (1989), p.178),[51] (Coles, vol.341 (1989), p.7),[52] (Danchin, vol.334 (1988), p.286),[53] (Doublet-Stewart, vol.335 (1988), p.200),[54] (Dunthorn, vol.335 (1988), p.664),[55] (Escribano, vol.334 (1988), p.376),[56] (Fierz, vol.334 (1988), p.286),[57] (Findlay, vol.335 (1988), p.292),[58] (Fisher, vol.335 (1988), p.292),[59] (Friedjung, vol.334 (1988), p.646),[60] (Gaylarde, vol.334 (1988), p.375),[61] (Gibson, vol.335 (1988), p.200),[62] (Gillman, vol.335 (1988), p.292),[63] (Glick, vol.334 (1988), p.376),[64] (Grimwade, vol.335 (1988), p.292),[65] (Hirst, vol.366 (1993), pp.525–7),[66] (Johnson, vol.335 (1988), p.392),[67] (Jonas, vol.335 (1988), p.292),[68] (Lane, vol.335 (1988), p.109),[69] (Lasters, vol.334 (1988), p.285),[70] (Lipowicz, vol.335 (1988), p.109),[71] (Maddox, vol.334 (1988), p.287),[72] (Maddox, vol.335 (1988), p.760),[73] (Metzger, vol.334 (1988), p.375),[74] (Miller, vol.340 (1989), p.498),[75] (Neville, vol.335 (1988), p.200),[76] (Nisonoff, vol.334 (1988), p.286),[77] (Opitz, vol.334 (1988), p.286),[78] (Petsko, vol.335 (1988), p.109),[79] (Plasterk, vol.334 (1988), p.285),[80] (Reilly, vol.334 (1988), p.285),[81] (Rothaupt, vol.335 (1988), p.758),[82] (Schilling, vol.335 (1988), p.584),[83] (Scott, vol.335 (1988), p.292),[84] (Seagrave, vol.334 (1988), p.559),[85] (Shakib, vol.335 (1988), p.664),[86] (Shoup, vol.335 (1988), p.664),[87] (Snell, vol.334 (1988), p.559),[88] (Stanworth, vol.335 (1988), p.392),[89] (Suslick, vol.334 (1988), p.375),[90] (Taylor, vol.335 (1988), p.200),[91] (Timmerman, vol.352 (1991), p.751),[92] (Van Valen, vol.335 (1988), p.664),[93] (Von Hahn, vol.335 (1988), p.664)[94]

New Biologist (Martin, vol.3 (1991), pp.409–11)[95]

New Scientist (Anonymous, 8 Sept. (1988), p.33),[96] (Cherfas, 5 Nov. (1988), p.25),[97] (Concar, 16 Mar. (1991), p.10),[98] (Patel, 23 Oct. (1993), p.10),[99] (Vines, 14 July (1988), p.39),[100] (Vines, 4 Aug. (1988), p.30–1)[101]

New-Zealand Medical Journal (Welch, vol.102 (1989), p.202)[102]

Quarterly Journal of the Royal Astronomical Society (De Jager, vol.31 (1990), p.31)[103]

La Recherche (Anonymous, vol.19 (1988), p.1005),[104] (Anonymous, vol.19 (1988), p.1149),[105] (Anonymous, vol.23 (1992), p.1224),[106] (Ourisson, vol.24 (1993), pp.1014–15)[107]

Science (Benveniste, vol.241 (1988), p.1028),[108] (Dickson, vol.245 (1989), p.248),[109] (Maddox, vol.241 (1988), pp.1585–6),[110] (Pool, vol.241 (1988), p.407),[111] (Pool, vol.241 (1988), p.658),[112] (Relman, vol.242 (1988), p.348)[113]

The Scientist (Benveniste, vol.2 (1988), p.10),[114] (Dixon, 5 Sept. (1988), p.1),[115] (Dixon, 5 Sept. (1988), p.5),[116] (Dorozynski, 5 Sept. (1988), p.4),[117] (Garfield, 5 Sept. (1988), p.12)[118]

Science Progress (Berezin, vol.74 (1990), p.495)[119]
Search (Trajstman, vol.20 (1989), p.14)[120]
Sourthern Medical Journal (Frenkel, vol.82 (1989), pp.1195–6)[121]
Thermochemica Acta (Labadie, vol.162 (1990), p.445)[122]
Thorax (Lane, vol.46 (1991), pp.787–97)[123]
Veterinary Record (Anonymous, 13 Aug. (1988), p.1)[124]

Thematic Classification of the Content of the Articles

Suggestions about artefacts

The suggestion published in French was analysed in Chapter 6 in the section entitled 'Scientific harassment'. The seven suggestions published by *Nature* have been analysed in Appendix 6c.

Reactions to censorship

The word 'censorship' was never used. The positions vary from severe criticism of the attitude of *Nature* (19, 20, 35, 60, 91) to a Macarthyist position where authors suggest that fraud squads should come to research laboratories without warning (63, 76).

Suspicions of fraud

This theme was analysed in the second part of the text of Chapter 7. Information about a specific suspicion of fraud was also given in Appendix 7a.

Homoeopathy

Most authors used the link between high dilution experiments and homoeopathy to discredit these experiments. The arguments used include references to financing (20, 21, 49, 72) or warnings against the dangers of making homoeopathic medicine appear respectable (13, 23, 36, 80). However, some authors point out that observations concerning homoeopathy represent a scientific challenge deserving more than sarcasm (62, 81).

Methodology

Practically all authors simply reproduce the arguments of Maddox, Randi and Stewart, with a special emphasis on inadequate statistical control. The issue of statistical variations of basophil counts has been examined in Chapter 6 and Appendix 6a.

Failures to duplicate high dilution experiments

Failures to duplicate high dilution experiments were analysed in Chapter 6, the most recent example being analysed in Appendix 6b. A synthetic view of high dilution experiments was presented in Appendix 6d.

Irony and sarcasms

The issue was analysed in the second part of Chapter 7. In particular, a list of unusual titles was given, followed by the text of three cartoons.

Theoretical critiques

The low level of scientific criticisms was illustrated at the end of Chapter 7 by a long list of quotations.

NOTES

Introduction

1 Born, *Structure Atomique de la Matière,* Armand Colin.

Chapter 1

1 When told of Benveniste's high dilution experiments using hearts, a scientist stated: 'You don't change the laws of physics with a heartbeat!'

2 A publication in a Canadian journal refers to the article of Del Giudice, Preparata and Vitellio without any comment. This article is also quoted in a Russian periodical which I did not consult.

Chapter 3

1 After 6 months, glass phials were replaced by plastic test tubes. Glass phials make a more dramatic impression but are harder to use than standard test tubes, being difficult to prepare in a sterile manner, and they have to be broken in order to test their contents.

Chapter 4

1 Physiological serum is prepared by adding the proper amount of salt to avoid the bursting of living cells.

Chapter 5

1 According to the first report published by Maddox, his visit ended on Friday 8 July. This date is also mentioned in another passage of the paper quoted here. The inconsistency can be seen either as an indication of a lack of care or as a Freudian slip revealing an uneasiness about rejecting 200 experiments on the basis of only two experiments. There were other examples of such lack of care in his report.

2 In the absence of any decision concerning his article, Benveniste had sent a fax to ask when he would get an answer, in order to decide whether he should submit his article elsewhere.

3 Quoted by Arthur Koestler in *The Sleepwalkers.*

4 Emphasis in the original.

Chapter 6

1 On another occasion, the demonstration of high dilution effects in a blind experiment supervised by hostile observers led to Elisabeth Davenas being called a 'witch.'

2 After grudgingly conceding that one of the last three experiments contained no information because the basophils were insensitive, Maddox forgot that fact in a later article, where he mentioned the results of 'three blind experiments'.

3 Even for a given type of chemical, the authors used three different types of blood sample, one for each dilution range.

Chapter 7

1 Anon. 'Amadeo Avogadro meets IgE', *The Lancet,* vol.ii (1988), p.117.

2 Maddox, J., Randi, J. and Stewart, W. W. 'High dilution experiments a delusion', *Nature*, vol.334 (1988), pp.287–90.

3 Benveniste, J. 'Dr Benveniste replies', *Nature*, vol.334 (1988), p.291.

4 When words are reproduced several times, it is because they were actually repeated in the various articles listed in Appendix 7b.

Chapter 8

1 Schiff, Michel. *L'Homme Occulté: le Citoyen face au Scientifique,* Paris, Editions Ouvrières, 1992.

Conclusion

1 Schiff, Michel. *L'Homme Occulté: le Citoyen face au Scientifique,* Paris, Editions Ouvrières, 1992, pp.68–78.

INDEX